95p

West African Nature Handbooks

Large Mammals of West Africa

D. C. D. Happold

Longman

Longman Group Limited
London

Associated companies, branches and representatives throughout the world

First published 1973

ISBN 0 582 60428 1

Printed in Singapore
by Singapore Offset (Pte) Ltd.

Contents

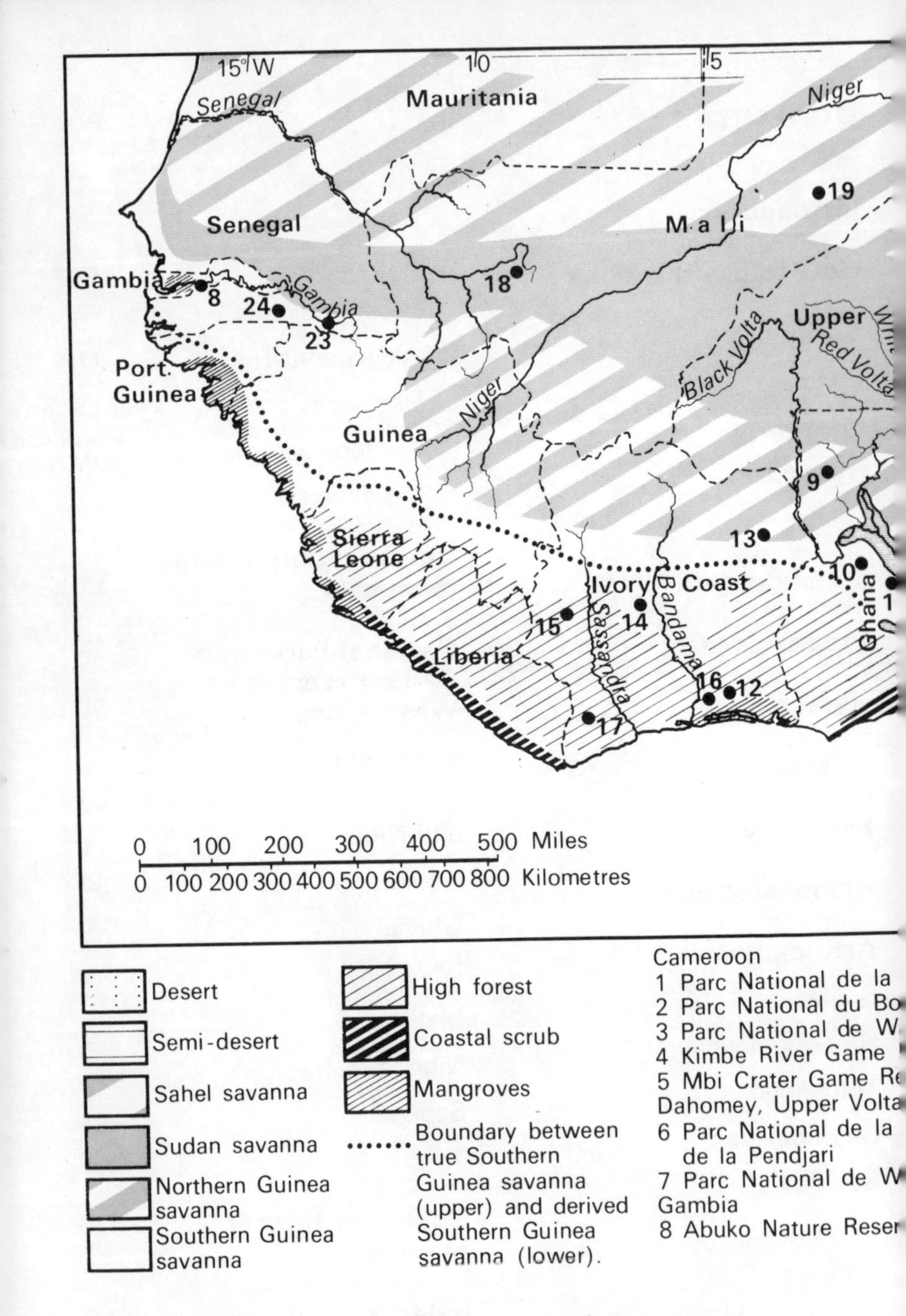

Fig A West Africa, showing vegetation zones and game reserves

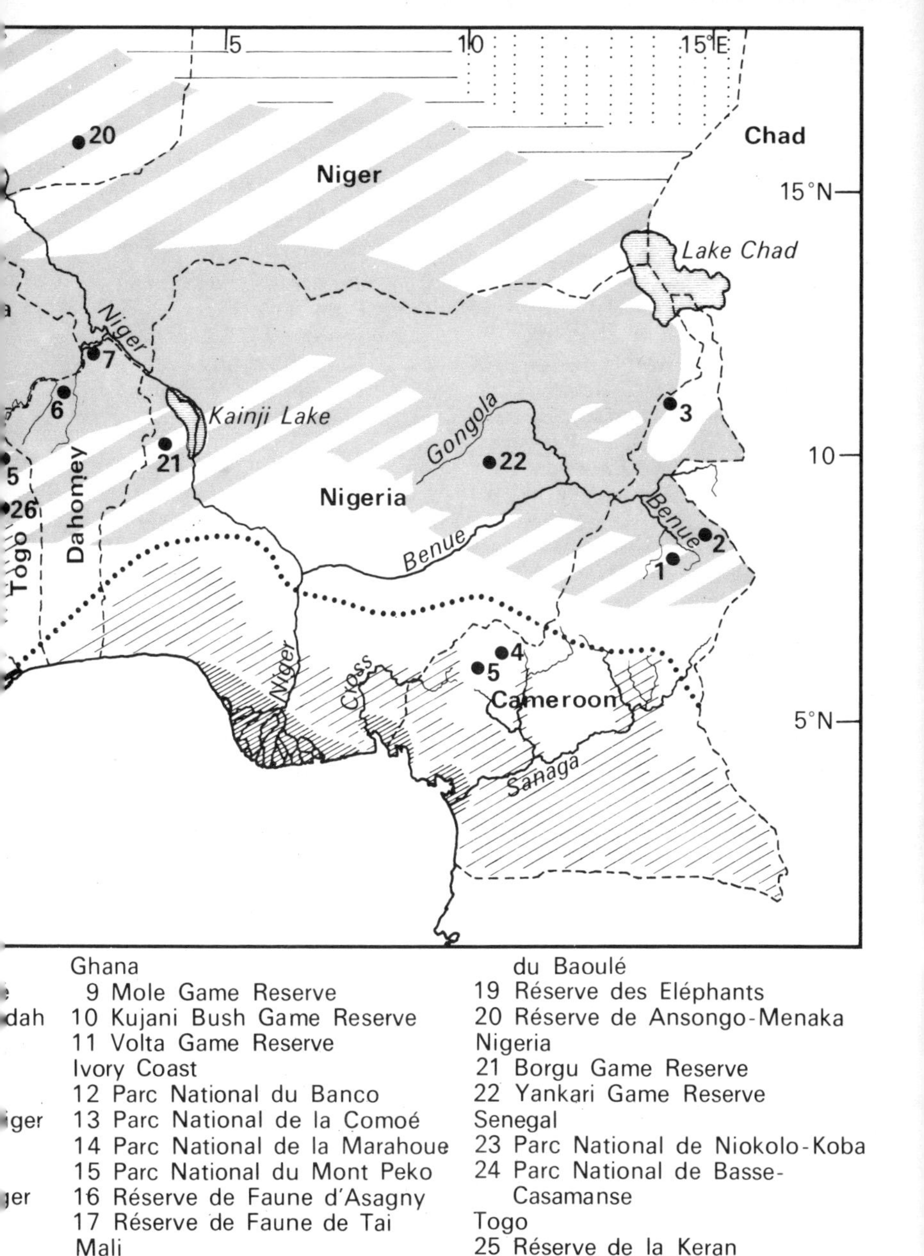

dah
iger
ger

Ghana
9 Mole Game Reserve
10 Kujani Bush Game Reserve
11 Volta Game Reserve
Ivory Coast
12 Parc National du Banco
13 Parc National de la Comoé
14 Parc National de la Marahoue
15 Parc National du Mont Peko
16 Réserve de Faune d'Asagny
17 Réserve de Faune de Tai
Mali
18 Parc National de la Boucle du Baoulé
19 Réserve des Eléphants
20 Réserve de Ansongo-Menaka
Nigeria
21 Borgu Game Reserve
22 Yankari Game Reserve
Senegal
23 Parc National de Niokolo-Koba
24 Parc National de Basse-Casamanse
Togo
25 Réserve de la Keran
26 Réserve de Malfacassa et Fazao

Acknowledgements

Many people have helped me collect the information for this book, and have read through all or parts of the manuscript. I am particularly grateful to the following for their help: Mr F. Adam, Mr W. H. F. Ansell, Mr E. Asibey, Mr Gilbert S. Child, Dr G. B. Corbet, Dr R. M. Cross, Mr J. Dubreuil, Mr A. R. Dupuy, Dr R. F. Ewer, Dr P. Grubb, Mr J. Henshaw, Mrs S. Jeffrey, Dr D. Kock, Dr F. Petter, and Dr J. Roche. Professor C. F. Hoffman gave much assistance on the West African names, and Dr J. Owen sent details on her captive Royal Antelope.

The section on reserved areas has been compiled from many sources, and I am grateful to the following individuals and organisations for sending information: Cameroon (East) – Mr J. Dubreuil; Cameroon (West) – Directorate of Forest Services, Buea; Dahomey, Upper Volta and Niger Republic – Mr J. Raynaud, Mr L. Blancou; Gambia – Mr E. Brewer; Ghana – Mr E. Asibey, Mr Gilbert S. Child; Ivory Coast – Department des Eaux et Forets, Mr C. Geerling; Nigeria – Mr J. Jia, Mr Gilbert S. Child, Mr J. Henshaw; Mali – Mr L. Macher, Department des Eaux et Forets; Senegal – Mr A. R. Dupuy; Togo – Department des Eaux et Forets; Mauritania – Dr J. M. Alonso.

I am especially grateful to my wife, Meredith, for the assistance she has given me throughout the preparation of this book.

Photographs

The publishers are grateful to the following for permission to reproduce photographs:
Antwerp Zoo (Photo V. Six): 27; G. Cansdale: 53, 64; Bruce Coleman Ltd: (D. Bartlett), 17, 18, 23, 29; (Jane Burton) 25, 28; (B. Campbell) 9, 10; (L. R. Dawson) 22; (N. Myers) 21, 56; (G. D. Plage) 13; Professor Dragesco: 65; A. R. Dupuy: 38; C. Geerling: 30, 32, 35, 40, 55, 58, 66; D. Happold: 11, 14, 29, 33, 36, 37, 44, 46, 48, 59, 60, 63, 69; Jacana Press: (R. Bousqult) 61; (A. R. Devez) 34; (J. L. S. Dubois) 2; (Frédéric) 42; (B. Fritz) 67, 68; (G. Haüsle) 4; (B. Mallet) 70; (F. Petter) 3, 16; (A. Visage) 5, 41; Sonia Jeffrey; 26, 51; R. H. Kemp: 1, 6, 20, 24, 54, 62; Zoological Society of London: 7, 31.

Drawings by Edward Osmond

Introduction

In 1960, A. H. Booth wrote *Small Mammals of West Africa* which described the insectivores and bats, monkeys, rodents and hares, pangolins, genets, civets and mongooses of West Africa. As he wrote in his Introduction, some of the mammals described are large but were included since their nearest relatives are small. This book describes the large mammals of West Africa but also includes some small ones which were not described in Booth's book since their relations are large.

Because of the bewildering numbers of species of small mammals, especially bats and rodents, it was not possible to include all of them in *Small Mammals of West Africa.* As the larger mammals and their relations are less numerous, I have been able to describe in this book all the 70 species which live in West Africa.

For the purposes of this book, the geographical limits of West Africa (Fig A) are:

Northern boundary:	18°N. latitude.
Eastern boundary:	eastern border of Niger Republic and Cameroon.
Southern and western boundaries:	southern border of Cameroon, and the sea coast from Cameroon to Mauritania at 18°N latitude.

During the last thirty years, several books in French have been published on the mammals of various parts of West Africa. These referred to the former Afrique Occidentale Française and none of them dealt with the mammals of the English speaking countries of West Africa. Most of these books are now unobtainable, are out of date since the A.O.F. no longer exists, and are of little value to the English-speaking people of West Africa. Checklists and some observations have been published in English but these are mostly in scientific journals and are not available for general reading.

During the last few years, many people in West Africa have shown an increasing awareness of the large mammals of their country. The numbers of many of these mammals have been greatly reduced in recent years, and regions which had large mammal populations in the past now have practically none. This destruction can be stopped only by an appreciation of the value of these mammals, and practical steps to ensure their survival. I hope that this book will stimulate interest and enthusiasm, and that this will lead to an appreciation of the value, beauty and importance of these mammals.

The first section of this book begins with a few instructions on how to use the book, followed by descriptions of the orders, families and species. The second section describes the principal conservation areas in West Africa where many of the species may be seen in their natural environment. Finally, there is a short list of additional books for the reader who is interested in learning more about the mammals of this part of Africa.

How to use this book

The mammals in this book are divided into seven orders and fourteen families; one family, the Bovidae, is divided further into nine sub-families*. When identifying a mammal, it is necessary to decide the order, then the family, and finally the species. A key is not provided since it will be fairly obvious to which order and family a mammal belongs.

The seven orders are:

Order	*Number of families*	*Number of species*
Carnivora – the carnivores	4	22
Tubulidentata – the aardvark	1	1
Proboscidea – the elephant	1	1
Hyracoidea – the hyraxes	1	2
Sirenia – the manatee	1	1
Perissodactyla – the rhinoceros	1	1
Artiodactyla – the antelopes and their relations	5	42

Domestic species are not described, but they are mentioned in the appropriate order and family.

All the species in these orders which occur in West Africa are described in the following pages; the only exception is one family of the Carnivora, the Viverridae (mongooses, genets and civets), which is described in *Small Mammals of West Africa*.
Information which is common to all members of a family, or order, is not repeated in the descriptions of each species. So it is necessary to read the order and family notes for each species to obtain the full information about a particular species.

The following information is given for each species.

Name

English and scientific names are given, followed by names in other languages.

F. = French
G. = German
H. = Hausa
T. = Twi
W. = Wolof
FU. = Fula (Fulani)
Y. = Yoruba
D. = Dyula
B. = Bambara
S. = Songhai

Between them, these eight local languages are spoken over a large area of West Africa. It has not been possible to include the many dialects, or languages, which are confined to small regions of West Africa.

Identification

Measurements give an indication of the size and proportions of each species; these measurements are average ones for adult individuals. If possible, the measurements are taken from specimens in West Africa since, for many species, there are differences between West African individuals and individuals of the same species from other parts of Africa. Unless otherwise stated lengths are in centimetres (cm) followed by the equivalent length in inches (in), and weights are in kilogrammes (kg) followed by the equivalent in pounds (lb).

SH = Shoulder height, from the ground to the top of the shoulder.
HB = Head and body length, from the tip of the nose to the

* A family name ends in -idae, e.g. Bovidae; and a subfamily name ends in -inae, e.g. Tragelaphinae.

base of the tail along the mid-dorsal line.

T = Tail length, from the base of the tail to the tip of the tail, excluding the terminal hairs (if any).

TL = Total length, from the tip of the nose to the end of the tail (manatee only).

HL = Horn length, from the base of the horn to the tip of the horn along the front edge.

WT = Weight of adult animal.

The description of the species gives the principal identification features such as colour, pattern, proportions, and other details useful for identification in the field. For two similar species, features are given which can be used to distinguish them.

Distribution*

This is recorded in two parts:

(a) In West Africa. It is not possible, at the present level of our knowledge, to draw detailed maps showing the geographical distribution of the species. Instead, the distribution is recorded by vegetation zones (since these determine to a large extent the distribution of mammals) and by country. If the distribution is discontinuous, each country where the species is recorded is stated; but for a widespread species, only the limits are recorded, e.g. 'Guinea savanna from Senegal to N. Cameroon' means that the species is found in the Guinea savanna in all countries from Senegal in the west to Cameroon in the east.

(b) In other places outside West Africa. This shows whether the species is widespread or rather limited in distribution. For some species, it shows that the species is found only in West Africa.

Habits

The information given in this section is mainly about the ecology of each species – where it lives, how it lives, what it eats, and some of the interesting facts about the species which can be easily seen when watching individuals in their natural habitat. There are many gaps in our knowledge about the habits of most of the species described in this book, and consequently there is very little to say about some species. Detailed studies in their natural environment are needed for many West African mammals.

Status

The status refers to West Africa only, although it may be applicable to the other parts of the range of a species as well. The terms 'rare', 'common', and 'abundant' are very subjective, and again reflect a lack of detailed information. But it is true that many of these large mammals are much rarer now than they were previously due to poaching and over-hunting by local inhabitants. Since the range of each species is determined by its ecological requirements, the status refers only to the abundance of the species within its possible geographical range, or its specific habitat.

*The species of large mammals which have been recorded from each West African country are listed by Happold (1972). See books for further reading, p. 101.

Carnivora

The Carnivora are the most successful order of meat-eating mammals. Most of their special characteristics are related to eating meat: well developed canine teeth for striking and holding the prey, cheek teeth for cutting and shearing meat, well developed cheek muscles for crushing meat and bones, pointed claws for catching and killing the prey (in the Felidae and some Viverridae), and a simple uncomplicated digestive system. The prey is caught by running (hunting dogs, cheetahs), or by stealth (many cats). Most carnivores are terrestrial, but some (otters) are semi-aquatic. Some supplement their own kills by extensive scavenging (hyaenas, jackals). Carnivores are found in high forest, savanna, semi-desert, and aquatic habitats.

Carnivores play an important ecological role by preventing their prey species becoming too abundant and ultimately destroying the habitat. They also help to maintain healthy populations of their prey by killing the old, injured and diseased. Consequently these predator species should be protected, and not killed as so many people incorrectly suggest.

There are five families in Africa:
Canidae – foxes, hunting dogs.
Mustelidae – weasels, ratel, otters.
Viverridae – civets, genets, mongooses (described in *Small Mammals of West Africa* by A. H. Booth).
Hyaenidae – hyaenas.
Felidae – cats.

Canidae

The Canidae are medium-sized mammals with an elongated muzzle, pointed or rounded ears, and a bushy tail. The legs are relatively long with non-retractable claws (cf Felidae). The colour of the hair is mostly uniform all over, except in the hunting dogs which have a pattern of black, white and brown patches. The prey varies from insects and rodents (foxes) to large antelopes (hunting dogs). They live and hunt singly (foxes) or in packs (hunting dogs). The prey is caught by chasing or running, sometimes over long distances. Members of this family are found in savanna, semi-desert and desert; none of them live in forest.

There are three sorts of canids in West Africa: jackals, foxes, and hunting dogs.

1 Common Jackal

Canis aureus
F. Chacal commun, G. Goldschakal, H. Dila, W. Tili, FU. Sundu, S. Nzongo.

Identification SH 41 cm (16 in), HB 66 cm (26 in), T 33 cm (13 in), WT $4\frac{1}{2}$ kg (10 lb).
A medium-sized dog-like animal with a bushy tail and no special distinguishing marks or patterns. Its fur is sandy to sandy-rufous, darker on the back due to black tips to the hairs. It has large pointed ears, often rufous on the back. Its flanks, legs and underparts are paler than the back without any black-tipped hairs. The tail is bushy and coarse with black tip.

Distribution (a) Sahel and Sudan savanna and semi-desert from Senegal to Niger. This species has a more northerly distribution than the side-striped jackal. (b) Sahel savanna and semi-desert of North

Africa extending to parts of East Africa, southern Europe, Asia and India.

Habits Common jackals are mainly nocturnal, but on cool days they may be active in the early morning and evening. Jackals make their own burrows where they rest during the day. At night they move about singly or in pairs. They feed on rodents and insects which they catch themselves, or on other animals killed by large carnivores. In some places jackals will search for food in rubbish dumps near villages. The call is a prolonged scream. Several young are born after a gestation period of about two months.

Status Fairly common in suitable habitats.

2 Side-striped jackal

Canis adustus
F. Chacal à flancs rayés,
G. Streifenschakal, H. Dila.

Identification SH 41 cm (16 in), HB 63–81 cm (25–32 in), T 36–38 cm (14–15 in), WT $4\frac{1}{2}$–7 kg (10–15 lb). A medium-sized dog-like animal. The general colour is a greyish-rufous, and the hairs on the back have alternating black and white bands to form a speckled pattern. The flanks and underparts are paler or white. There is a black line along the flank from the shoulder to rump with an indistinct white line on its upper margin. The ears are small, grey at the back and the tail is long and bushy, sometimes black at the tip. It is distinguished from the common jackal by the colour of the back of the ears and the side stripe.

Distribution (a) Guinea, and occasionally Sudan, savanna from Senegal to Nigeria and N. Cameroon. This jackal has a more southerly distribution than the common jackal. (b) Open savanna and mountains throughout most of Africa south of the Sahara.

Habits This species is similar to the common jackal. They are rather shy and therefore seen less frequently. The call is a single, short yap. There are 3–7 young born after a gestation period of about two months.

Status Not often seen. Generally uncommon, although common in some localities, e.g. Senegal.

3 Rüppell's fox

Vulpes rueppelli
F. Renard du desert, G. Sandfuchs.

Identification SH 25 cm (10 in), HB 41–51 cm (16–20 in), T 30 cm (12 in), WT $2\frac{1}{2}$–$3\frac{1}{2}$ kg (6–8 lb). A small, long-bodied animal with a bushy, white-tipped tail. The general colour is sandy, but the back is rufous and each hair has a white tip to give a peppered effect. The head is sandy with rufous around the eyes extending to the muzzle. The ears are white inside and deep rufous at the back. The flanks are pale with some black-tipped hairs, the underparts white. The soles of the feet have dense fur.

Distribution (a) Desert and semi-desert from Mauritania to Niger. (b) Desert and semi-desert of N. Africa extending to Arabia, Iran and Afghanistan.

Habits Very little is known about

these foxes. They are gregarious and probably live in colonies. They feed mainly on insects.

Status Probably very rare.

4 Pale fox

Vulpes pallida
F. Renard blond de sable, renard pâle, G. Blassfuchs, H. Yanyawa, FU. Doldolnde (?).

Identification SH 25 cm (10 in), HB 46–51 cm (18–20 in), T 25 cm (10 in), WT 2½ kg (6 lb). A small fox similar in size to Rüppell's fox. It has dense sandy-coloured fur, with some black or brown-tipped hairs on the back, a pointed muzzle, and ears that are rounded at the tip and white inside. The underparts are pale, sometimes white. The tail is similar in colour to the back with a black tip. It is distinguished from Rüppell's fox by the black-tipped tail, and from fennec fox by its larger size and smaller rounded ears.

Distribution (a) Semi-desert and Sahel savanna from Senegal to Niger. Sometimes in Sudan and Guinea savanna. (b) Semi-desert and Sahel savanna of northern Africa.

Habits Pale foxes are nocturnal, emerging from their burrows in the sand at dusk. These burrows may be up to 8 or 10 feet below the surface. Pale foxes often live in family groups, but usually hunt singly. They feed on rodents, lizards and birds.

There are 3–4 young in a litter.

Status Distribution patchy, but may be quite common in suitable sandy areas.

5 Fennec fox

Fennecus zerda
F. Fennec, G. Wüstenfuchs.

Identification SH 20 cm (8 in), HB 41 cm (16 in), T 23 cm (9 in). A very small, pale-coloured fox, with a long woolly coat, sandy-cream on the back and white underneath. The head is small, with a rounded muzzle and enormous triangular shaped ears. Sometimes there is a pale brown spot between the ears. The soles of the feet are densely furred. The tail is bushy, often black-tipped. This species is distinguished from Rüppell's fox and pale fox by its smaller size and larger ears.

Distribution (a) Deserts, especially sand dunes, from Mauritania to Niger; occasionally in semi-desert and Sahel savanna. (b) Deserts of North Africa and Arabia.

Habits Fennec foxes live in the sandy parts of deserts where they dig burrows in sand dunes. They live in groups of up to 10. They are nocturnal and feed on rodents and insects. For most of the time they do not require water, but drink when water is available. They make several sorts of vocal noises. The gestation period is nearly two months; there are 2–5 young.

Status Generally rare, but may be fairly common in some localities.

6 Hunting dog

Lycaon pictus
F. Cynhyène, Lycaon,
G. Hyaenenhund, H. Kyarkeci,
T. Sakraman, W. Safandu,
FU. Boinaru, Safandu,
S. Tagachi-Gandiachi.

Identification SH 76 cm (30 in), HB 114–122 cm (45–48 in), T 38 cm (15 in), WT 23–32 kg (50–70 lb).
A large slender dog-like animal with long legs. It has a broad head, with large rounded ears. The body colour is an irregular pattern of black, brown, sandy-yellow and white patches, the actual pattern varying in individual animals. The tail has a white tip.

Distribution (a) Guinea and Sudan savanna from Senegal to North Cameroon. (b) All suitable savanna regions of Africa.

Habits Hunting dogs are one of the important predators of the African savannas. They hunt in packs of up to 40 individuals in the early morning or at dusk, travelling long distances looking for their prey. They eat almost anything they come across from rodents and hares to the larger antelopes. Hunting dogs are adapted for sustained high-speed running in pursuit of prey. Their prey is not stalked, but chased by the pack until it becomes exhausted. It is then attacked on the flanks, often disembowelled, and then torn apart. Until recently, hunting dogs were considered a menace, but it is now realised that they play an important role in maintaining healthy populations of their prey species.

Litters of up to 10 young are born in holes in the ground after a gestation period of about two months.

Status Uncommon in most of West Africa although common in some localities, e.g. Niokolo-Koba Park.

Mustelidae

The Mustelidae consist of several species which superficially look rather dissimilar. Generally they have a rather elongated body with short legs. The head is rounded, with a shorter muzzle than in the dog family. There are five toes on each foot, and the claws are non-retractable. The hair is either short and dense, or long, and ranges from a uniform colour (some otters) to very distinctive light and dark markings (ratel, some weasels). Mustelids are very agile (except the ratel), and feed on insects, rodents, fish (some otters), and honey (ratel). They are usually solitary animals which move with a scampering gait and the back arched. Others are semiaquatic and are more agile in water than on land. Most mustelids have a characteristic smell due to secretions from the anal scent gland.

There are three sorts of mustelids in West Africa: weasels, ratel, and otters.

7 Zorilla

Ictonyx striatus
F. Zorille Commun, G. Zorilla, Band-Iltis, H. Bodari.

Identification SH 10 cm (4 in), HB 30–36 cm (12–14 in), T 23 cm (9 in), WT 1½ kg (3 lb).
A small agile weasel with an elongated body, short legs and distinctive black and white markings. The small rounded head is black with three white patches: one on each side from base of ear to above the eye, and a median one on forehead between the eyes. The tips of the ears are white. The top of the head, neck, back and sides are white with three long black lines from neck to base of tail. The legs and under-parts are black; the tail long and bushy, mainly white but with black at base of hairs. The back is slightly arched when walking, and hairs on back are erectile. Distinguished from the Libyan striped weasel by its slightly larger size, shorter hair, and greater amount of black colouration.

Distribution (a) Sudan and sometimes Guinea savanna from Senegal to North Cameroon. (b) Savannas of Africa.

Habits Zorillas are small nocturnal carnivores which are sometimes seen in the daytime. They live in burrows or under rocks, and are usually solitary. These weasels run with a bounding gait, but can also climb and swim. The food is mainly rodents, birds, reptiles and eggs. Zorillas are ferocious animals for their size, and when annoyed they utter a high pitched scream and erect the hairs on the back. Like most members of the family, they have a characteristic odour, and the anal glands produce a horrible smelling liquid if the animal is attacked.

There are 2–3 young, usually born in a burrow.

Status Uncommon to rare.

8 Libyan striped weasel

Poecilictis libyca
F. Zorille de Libye,
G. Streifenwiesel.

Identification SH 8–10 cm (3–4 in), HB 25–30 cm (10–12 in), T 13–15 cm (5–6 in), WT ½–1 kg (1–2 lb).
A small carnivore with long soft hair with black and white patterning. The black head has a white band from the angle of the jaw, over the head above the eyes, to the opposite side. The back is white with several indistinct black lines, and the under-parts are black. The tail is white, often darker at the tip. The soles of feet, except the pads, are hairy. Distinguished from the zorilla by its smaller size, longer hair and larger amount of white on the back.

Distribution (a) Semi-desert, and Sahel and Sudan savannas from Senegal and Mauritania to Niger. (b) Similar areas in Tchad and Sudan

Habits Little is known about these rare animals. Their habits are probably similar to those of the zorilla, except that they are adapted to more arid conditions.

There are 1–3 young in a litter.

Status Probably rare.

9 Ratel, Honey badger

Mellivora capensis
F. Ratel, G. Ratel, Honigdachs,
T. Sisi-kwabrafo, D. Trogba.

Identification SH 25 cm (10 in), HB 61–76 cm (24–30 in), T 25 cm (10 in), WT 11 kg (25 lb).
A medium sized heavily built animal with a broad white back and

black underparts. The most noticeable character is the white dorsal surface beginning on the head above the eyes, and extending over the neck, back and sides, to the base of the tail. The snout, legs and underparts are black. The tail is bushy, mostly white with some black hairs, sometimes held erect. The fore legs and toes are bent inwards and there are long claws. Sometimes black individuals are found in the forest.

Distribution (a) Semi-desert and savannas of most of West Africa; occasionally in forest. (b) Savannas of Africa, extending to parts of Arabia, Turkestan and India.

Habits Ratels are nocturnal and are seen most frequently at dusk, although they are sometimes seen in the daytime. During the day they rest in burrows, and sometimes in abandoned aardvark holes. Although mainly terrestrial, they can climb trees. Usually solitary or in pairs. Their usual food is rodents, insects and eggs. They are called 'honey badgers' because of their fondness for honey, which they obtain by tearing open bee's nests with their powerful claws. There is an interesting association between honey badgers and honey guide birds; these birds are said to guide honey badgers to the nests of bees by their calls. Like other members of this family, ratels have anal glands and a characteristic smell. Ratels are extremely strong for their size. When young, they are friendly in captivity, but very temperamental as adults.

Status Rare; seldom seen because they are shy and nocturnal, but may be commoner than supposed.

10 Spotted-necked otter

Lutra maculicollis
F. Loutre à cou tacheté,
G. Krallen-Otter, D. Kwa-uru.

Identification SH 30 cm (12 in), HB 63 cm (25 in), T 51 cm (20 in), WT 7 kg (15 lb).
A sleek, elongated, aquatic animal. Generally it is coloured deep chocolate-brown with white or pale yellow patches or spots on throat, chest and underparts. Its fur is dense, soft and short. It has a broad rounded head with small ears and many whiskers on the muzzle. The limbs are short and thick; the feet webbed, with small claws on the digits (cf clawless otter). The tail is long and very thick at the base.

Distribution (a) Suitable permanent rivers throughout West Africa in forest and Guinea savanna. Lake Tchad. (b) Suitable permanent rivers and lakes in Africa south of the Sahara.

Habits Spotted-necked otters inhabit permanent rivers, some mountain streams, and lakes. Like all otters they are excellent swimmers, and the webbed feet and thick tail aid them in fast and accurate movement in water. They are found singly or in small groups. Spotted-necked otters feed on fish which they catch in the water. They are thought to be more aquatic than the clawless otter.

The 2–3 young, born after a gestation period of about two months, are paler in colour than the adults.

Status Uncertain, probably generally rare, but common in some localities.

11 Clawless otter

Aonyx capensis
F. Loutre à joues blanches,
G. Weisswangen-Otter, H. Karen Ruwa, FU. Rawandu ndiyam.

Identification SH 36 cm (14 in), HB 91 cm (36 in), T 56–71 cm (22–28 in), WT 13–18 kg (30–40 lb).
A large, dark brown otter with a white patch covering the chest,

throat and sides of the face below the eyes and ears. (The hairs in this region are brown at the base and white at the tip.) The rest of the body is brown, sometimes with white spots on the underparts. The fur is dense, soft and short, with long whiskers on muzzle. The ears are small and rounded. The legs are short, with small webs between the digits especially on the hind feet, and no claws (cf spotted-necked otter). The tail is thick at the base.

Distribution (a) Suitable permanent rivers, streams or swamps throughout West Africa in forest, Guinea and perhaps Sudan savanna. Lake Tchad. (b) Suitable permanent waterways in Africa south of the Sahara.

Habits It is thought that clawless otters are less aquatic than the spotted-necked otters since they feed mainly on crabs and molluscs, and sometimes on fish. The cheek teeth are adapted for crushing the hardened outer cases of the prey. They are mainly nocturnal, but sometimes sun themselves on rocks near the river. They are found singly or in family groups. These otters do not dig their own burrows, and when the young are born the mother uses a burrow abandoned by another animal. The call is a shrill piercing whistle.

The 2–5 young are born after a gestation period of about two months; they are paler in colour than the adults.

Status Uncertain, probably generally rare.

12 Cameroon otter

Paraonyx microdon
F. Loutre de Cameroun (?).

Identification SH ?, HB 76 cm (30 in), T ?, WT 20–27 kg (45–60 lb). (Not illustrated.)
A very large otter similar to, but heavier than, the clawless otter. Its general colour is dark brown with white markings on face and chest. The toes of the fore feet are partly webbed and those of the hind feet fully webbed. The claws are small and blunt.

Distribution (a) Certain rivers and marshes near Bamenda and Bipindi in central and southern Cameroon. (b) Not found elsewhere, although another species of *Paraonyx*, which may be conspecific with the Cameroon otter, occurs in the Congo basin.

Habits Very little is known about these otters. They are probably more terrestrial than the other otters, and they probably feed on amphibians and small land vertebrates. The teeth are not strong, and are incapable of crushing like those of the clawless otter.

There are 2–3 young in a litter.

Status Very rare.

Hyaenidae

The hyaenas are medium-sized or large carnivores. They are scavengers feeding on carrion, but they also kill their own prey. The teeth, jaws and cheek muscles are developed for crushing bones. The hind limbs are shorter than the fore limbs and the back slopes downwards towards the tail. The hair is short or long, and patterned with spots or stripes. The claws are non-retractile, and there is a well developed sense of smell. Although hyaenas look like dogs, they are more closely related to the cats.

Hyaenas are found in open country; they are usually solitary but the spotted hyaena sometimes live in packs.

There are two species in West Africa: the spotted hyaena and the striped hyaena.

13 Spotted hyaena

Crocuta crocuta
F. Hyène tachetée, G. Tüpfel Hyane, Gefleckte Hyäne, H. Kura, T. Pataku, FU. Føuru, Y. Kòríkò, Ikoókò, D. Suruku, S. Koro.

Identification SH 76–91 cm (30–36 in), HB 127–152 cm (50–60 in), T 25–30 cm (10–12 in), WT 45–54 kg (100–120 lb).
A large dog-like animal with a spotted coat, heavily and strongly built. Its general colour is dull grey to greyish-brown, with brown to blackish spots on the back, flanks and rump, sometimes indistinct. The head is large, rounded and powerful, the muzzle short, the ears rounded, the neck thick. The fore legs are longer than the hind legs so that the back slopes downwards to the base of the tail, which is short with a black bushy tip. The hair is coarse and woolly.

Distribution (a) Guinea and Sudan savannas from Senegal to Cameroon. (b) Throughout suitable savannas in Africa, including mountainous regions.

Habits Spotted hyaenas are mainly nocturnal, spending the daytime in burrows or under cover. They are scavengers feeding on the remains of prey killed by large carnivores, but will also catch and kill small prey, e.g. antelopes and hares. They kill large prey when hunting in packs. Their powerful jaws enable them to break and chew large bones. They are found singly or in groups of up to about eight animals. Spotted hyaenas are often aggressive and noisy; they have several vocal noises including a typical howl and 'laugh'. In some areas, hyaenas are bold and look for food close to human settlements.

The gestation period is about three and a half months, and there are 1–2 young. The young have no spots but are blackish in colour.

Status Common in suitable habitats.

14 Striped hyaena

Hyaena hyaena
F. Hyène striée, Hyène rayée, G. Streifenhyäne, H. Sayaki, Kure-Kure, W. Bøuki, FU. Yokoldu, S. Chabo-diano.

Identification SH 69–76 cm (27–30 in), HB 102–114 cm (40–45 in), T 30–46 cm (12–18 in), WT 36–54 kg (80–120 lb).
A medium dog-like animal with back sloping downwards towards the tail, and black vertical stripes on the sides. Its general colour is pale grey or beige, and it has a black patch on the throat, 5–7 indistinct vertical stripes on the flanks and clearer black stripes on the fore and hind legs. The head is roundish with a pointed muzzle (cf spotted hyaena) and pointed ears. It has a mane, which can be held erect, along the mid-dorsal line. The tail is bushy and the hair generally coarse and long.

Distribution (a) Sudan and Sahel savannas from Senegal to N. Cameroon. (b) Similar habitats in northern Africa, eastwards to Somalia, Kenya, Uganda and north Tanzania. Asia Minor, Turkestan, Afghanistan and India.

Habits Striped hyaenas have a more northerly distribution than spotted hyaenas, and they are less noisy and aggressive. They are nocturnal, spending the day in holes in the ground, e.g. abandoned aardvark holes. They are found singly or in pairs. They are mostly scavengers, but will eat anything that is available. Like the spotted hyaena, they can break and chew bones. Although they live in a dry area, they need to drink regularly and consequently they range over large areas looking for food and water.

There are 2–4 young, born after a gestation period of about two months. The young are the same colour as the adults.

Status Generally rare.

Felidae

The Felidae are the most specialised of the carnivores. They are large (lion) or small (sand cat), and are plain in colour or spotted. The head is rounded with a short muzzle and small rounded ears (except in caracal and sand cat), the legs are relatively thick with retractile pointed claws (except cheetah), and the tail is usually long. Cats always kill their own prey either by stealth (leopard), stealth and running (lion), or running (cheetah). They are extremely strong and can be dangerous if cornered or wounded. Most of the felids are terrestrial, but some of them are good climbers (leopard).

They live either singly or in pairs, or in groups, and are distributed over a wide range of habitats: forest (African golden cat), savannas (most cats), semi-deserts and desert (sand cat).

The vocal noises of felids are comparatively quiet. All felids purr when contented and hiss when frightened or annoyed. Lions roar, usually at night.

Tigers do not occur in Africa; they are found only in certain forests of Asia, Sumatra and Java.

15 African wild cat

Felis libyca
F. Chat sauvage de brousse, Chat sauvage d'Afrique, G. Afrikanische Wildkatze, H. Muzurun Daji, FU. Faturu ladde.

Identification SH 36 cm (14 in), HB 51 cm (20 in), T 33 cm (13 in), WT $5\frac{1}{2}$–$6\frac{1}{2}$ kg (12–14 lb).
Rather similar to a domestic cat in size and proportions, the wild cat is grey to rufous in colour, usually darker on the mid-dorsal line, with indistinct stripes or spots on the flanks. The backs of the ears are rufous in colour. There are black hairs around the pads of the feet. The underparts are paler than the back. The tail is darker, often with black rings near the tip, and shorter than in domestic cats. Southern specimens are darker than those in semi-desert areas.

Distribution (a) Desert, semi-desert and savanna regions of West Africa. (b) Savanna, semi-desert and mountainous regions of Africa, Asia Minor and S. Asia.

Habits Little is known about wild cats because they are nocturnal and secretive. During the day they

15

remain hidden in thick bushy areas or among boulders. Their food consists of birds, rodents, lizards and occasionally fruit. The gestation period is 56 days, and there are 2–5 young. Wild cats may interbreed with feral domestic cats.

Status Like other nocturnal members of this family, it is difficult to assess its status. Although the species has a wide range, its distribution is patchy; it is usually considered to be rare, but may be commoner than realised.

16 Sand cat

Felis margarita
F. Chat des sables, G. Saharakatze.

Identification SH 25 cm (10 in), HB 43 cm (17 in), T 13 cm (5 in), WT 2½–4 kg (6–9 lb).
A small, sandy-coloured cat, with black markings on tips and back of ears, upper parts of fore legs and tip of tail. The head is rather broad with long ears placed low on the side of the head. The fur is thick and soft and there are long greyish hairs on the soles of the feet.

Distribution (a) Semi-desert and desert areas from Mauritania to Niger. (b) Scattered desert localities of north Africa, Middle East, Arabia and southern USSR.

Habits Like other small desert animals, sand cats are nocturnal and remain underground in burrows during the day. Most burrows tend to be where sand dunes are numerous. Sand cats emerge at dusk to hunt for rodents, which are their main source of food. The long hairs on the soles of the feet probably aid movement over soft sand. There are 2–4 young; markings are darker in the young than the adults.

Status Very rare or rare.

17 Serval

Felis serval
F. Serval, G. Serval, H. Kwara, Kawun Damisa, T. Obata, Kotoku,

W. Babafal, FU. Njaluwa cirgu, D. Ngolongari.

Identification SH 56 cm (22 in), HB 71–89 cm (28–35 in), T 25 cm (10 in), WT $13\frac{1}{2}$–18 kg (30–40 lb). A medium-sized cat, pale with dark spots, the serval has long legs and a shortish tail. The coarse, woolly fur is pale beige or sandy with small black stripes on the back and irregular small spots on the sides. Underparts white with small black spots on belly and insides of legs. The head is proportionally small with two black bands from top of head to internal corner of eyes. The ears are rounded and oval, usually with two black stripes on the back. The tail is beige with black rings. The West African variety, sometimes called the servaline cat (*Felis brachyura*), has smaller spots and looks speckled. It is often confused with the leopard, a much larger, heavier animal with very distinct spots clustered together to form rosettes.

Distribution (a) Savanna and occasionally forest regions, including montane forest, from Senegal to Cameroon. (b) Savanna areas of Africa, except South Africa.

Habits Very little is known about servals, although they are quite well known by name. They are nocturnal, but may sometimes move about in the early morning and at dusk. During the daytime, they often remain in rocky areas. Their food consists of small antelopes and duikers, rodents and birds; some individuals living near human habitations have become very destructive to domestic poultry.

There are 2–4 young.

Status Probably not uncommon, but rarely seen because of their secretive, nocturnal way of life.

18 Caracal

Felis caracal
F. Caracal, G. Karakal, H. Metso, Dagiri, FU. Sowondu (?).

Identification SH 46 cm (18 in), HB 76 cm (30 in), T 30 cm (12 in), WT 12½–18 kg (28–40 lb).
A medium-sized pale cat with pointed, tasselled ears. The fur is thick and dense, varying in colour from beige to red-brown with very indistinct small spots in some individuals. The head is large and flattened, with a black stripe from the corner of the eye to the nose. The ears are long and pointed with a black tuft of hairs at the tip, and black on the back. The legs are the same colour as the body. The tail appears rather short. Black individuals have been recorded.

Distribution (a) Guinea and Sudan savanna from Senegal to Cameroon. May extend into derived savanna along northern edge of forest. (b) Most savanna regions of Africa, Arabia and India.

Habits Caracals are mostly nocturnal and prefer to live in rocky areas where they hide during the day. They are good climbers and very agile. Their food is mainly rodents, hares, birds, rock hyrax and perhaps small antelopes. When disturbed, caracals make a spitting or hissing noise. There are 2–4 young in a litter.

Status Rarely seen; probably uncommon.

19 African golden cat

Felis aurata
F. Chat doré, G. Goldkatze, T. Obata-kokoo, D. Bara, Kolukari.

Identification SH 51 cm (20 in), HB 76 cm (30 in), T 38–46 cm (15–18 in), WT 13½–16 kg (30–35 lb).
A medium-sized cat with short legs and a long tail. The fur is short and soft, very variable in colour from golden-red, to grey-brown, or brownish-black. The underparts are paler, almost white in some individuals, with small, irregular black spots, extending on to flanks and back in some specimens. The ears are roundish, black on the back, and the tail is long and similar in colour to the body.

Distribution (a) Forest regions from Senegal to Cameroon. May extend into gallery forests in Guinea savanna but have not been recorded in Nigeria. (b) Forests of Congo basin and Uganda, and highland forests of Kenya.

Habits Little is known about the habits of golden cats. They are shy and probably lead a life similar to that of leopards, resting in trees in the daytime and hunting at night. Golden cats prefer moist forests, especially close to rivers. They feed on small mammals, and domestic poultry if available.

Status Very rare, though probably not uncommon in Cameroon.

20 Lion

Panthera leo
F. Lion, G. Löwe, H. Zaki, T. Gyata, W. Gainde, FU. Bilade, Njagawu, Y. Kìnìún, D. Diara, B. Waraba, S. Gandihaya.

Identification SH 102 cm (40 in), HB 203–254 cm (80–100 in), T 89–102 cm (35–40 in), WT 136–

204 kg (300–450 lb).
A large, powerful cat. Its fur is short, silver-grey to ochre, usually darker on the head. The underparts are lighter. The head is very large, similar in colour to body. The ears are small and rounded, with black markings on the back. Male lions have a mane of long hair on their neck and shoulders, which is usually black, brown or rufous-yellow in colour. There is a small tuft of hairs on the elbow. The legs are thickset with large paws and the tail is rather long with a black tuft at the tip. The young are spotted until they are about a year old.

Distribution (a) Guinea and Sahel savanna from Senegal to Cameroon. Formerly extended northwards into what is now the Sahara desert. (b) Most savanna regions of Africa.

Habits Lions are probably the best known of the larger mammals of Africa. They are sociable animals living in family groups, or 'prides', of up to 25 individuals, but they are not tolerant of strange lions which are sometimes attacked. Prides have their own territories. Lions usually rest on the ground during the day, but in some areas they climb into trees. Lions usually hunt at night, either singly or in small groups. They approach their prey by stalking, and finally by a short fast run. The prey is killed by a throat bite and by breaking the neck, or by suffocation. Sometimes they ambush their prey, or when hunting in a pride they drive the prey towards other lions which are hiding out of sight. The lion is not capable of fast sustained running (cf cheetah), and it gives up if unable to catch its prey after a short run. The usual prey are antelopes, but the favoured prey species varies from place to place. Lions' senses of smell, sight and hearing are good. They have a characteristic roar, often followed by 4–8 grunts each diminishing in loudness.

The gestation period is about 105 days; there are 2–6 young.

Status Lions can exist only where there is a good population of prey species. Consequently they are rare in West Africa, except in protected areas.

21 Leopard

Panthera pardus
F. Panthere, Léopard, G. Leopard, H. Damisa, T. Osebo, W. Segue, FU. Dobbon, Cirgu, Y. Ekùn, Ògìdán, D. Warakalan, B. Waranikala, S. Maru.

Identification SH 71 cm (28 in), HB 127 cm (50 in), T 86 cm (34 in), WT 68–77 kg (150–170 lb).
A large, heavily built cat with characteristic spots. The fur is dense and soft. The background colour is buff or yellowish-brown, and on the back and sides there are clusters of 3–4 black spots which form rosettes. There are black spots on the head, legs and tail. The underparts are paler than the back, sometimes white, with irregular black spots. The head is broad with small rounded ears. The legs are thick and short. The very long tail is usually curled up slightly at the tip. There is considerable variation in the shape and arrangements of the rosettes and spots. Forest leopards are darker than savanna ones and black leopards have been recorded.

Distribution (a) Forest and savanna from Senegal to Cameroon. (b) Forest and forested savanna in Africa and Asia.

Habits Leopards are solitary and mainly nocturnal. During the day, they lie on branches of trees, and are difficult to see because the patterning of their fur breaks up the outline of the body. They stalk their prey without running. The prey is killed by a throat bite, and is often hauled into a tree out of reach of scavengers. Leopards are very powerful and cunning, and are excellent climbers. They prefer to remain in dense vegetation, and close to water since they usually drink each day. The prey is usually medium-sized antelopes; but they are also fond of baboons and consequently reduction of the leopard population often results in plagues of baboons. Leopards make a rough coughing sound.

The gestation period is about three months, and there are 2–3 young in a litter. The young are spotted like the adults.

Status The present distribution is patchy and it is generally assumed that leopards are commoner than realised. Leopards are killed for their skins and because they sometimes attack domestic stock. Their numbers are declining and they are probably rare in most localities.

22 Cheetah

Acinonyx jubatus
F. Guépard, G. Gepard, H. Argini, W. Tene, FU. Nyague-barode, Padawu, D. Kolokari, B. Kolokari.

Identification SH 76 cm (30 in), HB 127 cm (50 in), T 71 cm (28 in), WT 45–64 kg (100–140 lb).
A large, lightly built and very graceful cat with long thin legs, hollow back, and a long tail. The fur is beige or sandy with small irregular black spots or streaks (cf leopard). The head is relatively small, with a few black spots and a well defined black line from the corner of each eye to the mouth. The tail is similar in colour to the body, with black markings often forming rings. The claws, unlike those of other members of the cat family, are non-retractile. The young have smoky-grey woolly fur, with no markings.

Distribution (a) Guinea and Sudan savanna of West Africa. (b) Savannas of Africa, Arabia and Iran, and formerly India.

Habits Cheetahs are well known as the fastest terrestrial mammal, reaching a speed of about 95 km/h (60 mph). Living in open savanna areas, they are able to run down their prey, often over distances of up to 400 m. They hunt singly or in small groups at dawn or dusk, and they rely on their sight and speed to capture their prey, which are usually small antelopes. There is usually no stalking when hunting in open country; but sometimes cheetahs will lie in ambush until the prey has come within range for running it down. The prey is knocked over and killed by a throat bite, or it is knocked over in such a way that it breaks its neck.

The gestation period is about 95 days, and there are 2–4 young in a litter.

Status Cheetahs are rare throughout the whole of their range, and almost extinct in West Africa. These delightful animals require absolute protection.

Tubulidentata

This is a very specialised order of mammals which is adapted for eating ants and termites. The elongated snout, long tongue, strong limbs and claws are adaptations for breaking open the nests of termites and ants and obtaining the insects. The number of teeth are reduced since only crushing teeth at the back of the mouth are necessary for chewing. There is only one species, the aardvark.

23 Aardvark

Orycteropus afer
F. Oryctérope, G. Erdferkel, H. Dabgi, Daugi, T. Pantan, Wankyi, FU. Yendu, Y. Afèrìbòjǒ, D. Timba, S. Minei.

Identification SH 63 cm (25 in), HB 102–127 cm (40–50 in), T 61 cm (24 in), WT 59 kg (130 lb). A pig-sized animal with an arched back, long snout and ears, and small dark eyes. The greyish-pink skin is almost hairless, the body round and barrel-like. The legs are shortish and thick with long powerful claws. The tail is thick at the base, tapering towards tip. It carries its head close to the ground, and the tip of its tail trails on the ground.

Distribution (a) Guinea and Sahel savanna from Senegal to Cameroon. (b) Savanna regions of Africa south of the Sahara.

Habits The aardvarks or 'earth pigs' are solitary, nocturnal animals, and although they are quite common in some regions, they are rarely seen. During the day, aardvarks rest underground in burrows which they dig for themselves with their powerful legs and long claws. Often their presence is known only by these large burrows. Aardvarks break open the nests of termites with their claws, and then feed on the insects in the nests. The long sticky tongue

23

is extruded into the nests, and the insects are drawn into the mouth. The senses of hearing and smell are well developed, but sight is poor. Aardvarks have to move over considerable distances looking for nests, and their abandoned burrows are then used by other animals, e.g. warthogs, small carnivores and snakes.

The gestation period is about seven months; there is usually one young.

Status Their distribution is patchy, but they are probably fairly common in some areas. A supply of ants' and termites' nests is essential for their survival; their numbers have probably decreased in recent years.

Proboscidea

The two species of elephants, the African elephant and the Indian elephant, are the only survivors of what was once a very widespread and successful order. They are characterised by their enormous size, the elongation of the nose and upper lip to form a trunk, and the development of the third upper incisors to form tusks which grow throughout life. The bones are large and heavy; there is a relatively smaller amount of muscle compared with other mammals, but a lot of elastic connective tissue which helps to support the weight of the body. The legs are thick and stumpy, and the bones in the circular feet are supported by pads of cartilage. Despite its large size, an elephant can run very fast over short distances (up to 40 km/h (25 mph)).

Elephants are entirely vegetarian, feeding on leaves, shoots, fruits, bark, and roots. The trunk and tusks are used for obtaining food. The cheek teeth of the elephant are replaced several times during life; at any one time there are four teeth in use (one on each side in upper and lower jaws) and when one tooth is worn down it is replaced by a new one.

Elephants become sexually mature at about 15–20 years of age, and may live as long as 70 years.

24 African elephant

Loxodonta africana
F. Eléphant d'Afrique, G. Afrikanischer Elefant, H. Giwa, T. Esono, W. Niei, FU. Nyiwa, Y. Erin, D. Samatiatio, B. Simba, S. Tarkundei.

Identification SH 274–366 cm (9–12 ft), HB 550–762 cm (18–25 ft), T 91–122 cm (3–4 ft), WT 4070–6100 kg (4–6 tons).
This elephant is the largest terrestrial animal, and is too well known to need much description. Its skin is grey and wrinkled, with sparse coarse bristles. It uses its trunk for feeding, grooming, drinking, and fighting, and can move and flap its large ears, which are often slightly pointed at the base. It has a massive, sack-like body with thick, straight legs, its weight supported by elastic pads in the feet. There are four or five toes on the fore feet and four on the hind feet. The tail is relatively small with a tuft at the tip.

Two varieties or subspecies (originally classified as species) are recognisable, although there are many intergradations between the extremes. (1) *africana*, the savanna elephant; very large, with broad ears that are slightly pointed at base; SH 274–366 cm (9–12 ft). (2) *cyclotis*, the forest elephant; smaller than the savanna form with smaller ears that are rounded at the base; SH 214–274 cm (7–9 ft).

Distribution (a) Forest and savanna regions throughout West Africa. (b) Most of Africa south of the Sahara, including mountains; rare in South Africa; formerly present in N. Africa where it survived until early historical times.

Habits Elephants live in most habitats south of the Sahara. They are sociable animals, often living in herds of up to 50 animals. The family group is the mother, daughters and calves; these groups often join to form large herds. Bulls live in small groups and in small herds of up to about 12 individuals. Some old bulls are solitary. Elephants spend much of the day feeding, and they travel great distances while searching for their preferred food. They browse on leaves of trees, and less frequently graze on grass although in some localities (e.g. Borgu Game Reserve), grass forms the main part of the food during the rains. Food is grasped by the tip of the trunk and passed in to the mouth. The tusks are used to scrape bark from the sides of trees. Elephants often frequent salt licks, and they usually drink once each day. They will dig for water in dried up river beds. Pools and rivers are also used for bathing, and water is drawn up in the trunk and sprayed over the head and back. If the elephants are too numerous, they modify the vegetation and forested savanna can become almost treeless. Forest elephants live in small herds of about 2–10 animals; fallen fruits often form their main food.

Elephants make several noises: rumblings (made in the throat), trumpetings (often associated with aggression), and screamings.

The gestation period is about 22 months; there is one calf.

Status Elephants have become rare in recent years due to poaching and hunting for ivory, but they are still fairly numerous in some localities, especially in national parks and reserves, and some forest areas (e.g. Ghana).

Hyracoidea

The hyraxes are similar to domestic rabbits in size, but with a very short tail and short ears. There are four digits on the forefoot, and three on the hindfoot, and the soles of the feet are naked and well supplied with sweat glands so they are always moist. These specialised feet enable the hyrax to climb on steep smooth surfaces.

Hyraxes are herbivorous, feeding on fruits, shoots, and leaves. Their stomach is simple and they do not ruminate (see p. 27); the teeth show resemblances to those of ungulates, but the upper incisors are elongated to form small tusks which are triangular in section. Hyraxes are diurnal or nocturnal.

The young are very advanced at birth, with their eyes open, some teeth developed, and covered with hair. They can climb almost as well as adults.

Hyraxes have several special calls which are for alarm and for communication among members of a group.

There is one family, represented by two species in West Africa; these two are a good example of ecological separation between two closely related species:

	rock hyrax	tree hyrax
Distribution	savanna	forest
Habitat	rocks	trees
Activity	daytime (mainly)	nocturnal (mainly)
Social behaviour	gregarious	solitary, or small groups
Food	grasses, fruits	leaves, fruits
Dorsal gland	small	large

Hyraxes are thought to be related to the elephants and the Sirenia, even though they look more like rodents or rabbits.

25 Rock hyrax

Procavia ruficeps
F. Daman de rocher, G. Klippschliefer, H. Rema, Agwada, FU. Jama wuddere, D. Farabani, B. Kulu-Bani.

Identification SH 20–30 cm (8–12 in), HB 50 cm (20 in), WT 2½–4 kg (6–9 lb).
A small, rabbit-sized animal. The fur is soft and dense, grey to brown sometimes with patches of rufous. It has a slightly pointed nose, dark eyes and small ears held close to head. Its body is compact and low, so that it almost touches the ground.
It has no tail. The feet are flattened with large black pads. There is a patch of erectile hairs (surrounding the dorsal gland) on the back. The rock hyrax is distinguished from the tree hyrax by its brown colour, lack of white patch on back, and habitat.

Distribution (a) Rocky areas throughout savanna regions of West Africa from Senegal to Cameroon.
(b) Rocky areas throughout Africa, including desert regions, Arabia and the Middle East.

Habits Rock hyraxes are sociable animals living in colonies on rocky hills. They are well adapted for running up and down the steep sides of rocks, often using well defined trails in really steep places. Rock hyrax colonies are sometimes widely separated, and hyraxes may be able to cross savanna areas to colonise new rocky habitats. They are active in the early morning, afternoon and evening when they sun themselves and feed on fruits, roots, leaves and shoots. Sometimes, especially at night, they leave the rocky areas and feed on the fruits of surrounding trees. They shelter in crevices and cracks in the rocks, and these shelters are essential to their survival. Colonies of hyraxes may contain up to 50 or more animals. They have many communication sounds, including a shrill whistle which is the most frequently heard sound. Hyraxes usually have special defaecation places where mounds of droppings accumulate.

The gestation period is about seven months and there are 2–3 young in a litter.

Status Usually common in rocky areas which are sufficiently extensive.

26 Tree hyrax

Dendrohyrax dorsalis
F. Daman d'arbre, G. Baumschliefer, T. Owea, Y. Òfàfà, D. Ahoya.

Identification SH 30 cm (12 in), HB 38–50 cm (15–20 in), WT 2–2½ kg (4–6 lb).
Similar in appearance to rock hyrax, the tree hyrax has long very dark or black hair, soft in young animals but coarse in adults. A white or yellow patch in the middle of its back surrounds the dorsal gland. Distinguished from rock hyrax by the colour of this back patch, longer hair and habitat.

Distribution (a) Forest areas of West Africa from Sierra Leone to Cameroon. (b) Not found elsewhere, but a similar species *D. arboreus* occurs in forested regions of east, central and southern Africa.

Habits The tree hyrax is found in high forest. Like rock hyraxes they are good climbers, running up and down trunks and branches; but young hyraxes are not good climbers until several months of age. They are nocturnal, but sometimes feed in the early morning. Sometimes a tree hyrax will drop vertically to the ground from a low tree, and run quickly on the forest floor. Tree hyraxes are usually solitary, or live in small family parties. They feed on leaves, shoots and fruits. Because of their way of life, tree hyraxes are rarely seen and are best known by their call: an ascending scale of barks ending in 2 or 3 very loud barks. This noise, rather like that of a dog, has resulted in tree hyraxes being incorrectly called 'bush dogs'. Another local name, 'tree bear', is due to their bear-like appearance.

The gestation period is about seven months; there are 1–2 young, which are fully developed at birth. They are capable of feeding on soft shoots and are able to clamber about on branches and creepers.

Status Difficult to assess, but survival is dependent on large areas of natural forest with tall trees. They may be common in suitable areas but are probably declining due to the destruction of high forest. They can survive in farmland if there are sufficient trees.

Sirenia

The Sirenia, together with the whales, are the most aquatic of all the mammals. The body is large and elongated with practically no hair. The fore limbs are reduced in size and shaped like paddles and the hind limbs have been lost. The tail is a horizontally flattened fluke (as in whales). The head is rounded, with very small eyes, large upper lips covered by vibrissae (whiskers), and nasal openings on the tip of the snout. All these adaptations allow the sirenians to remain continually in the water, often completely submerged, and to feed on water weeds growing in the shallow water. They cannot move on land, They are vegetarians, have a three-chambered stomach, and the cheek teeth are replaced in a similar way to those of elephants.

Sirenia live mostly in estuaries and coastal lagoons, but some species spend all their lives in permanent rivers (e.g. the West African species).

The single young is born in the water, and is nursed by the mother who suckles it by holding the young between her fins.

There are four living species, one of them in West Africa.

27 West African manatee

Trichechus senegalensis
F. Lamantin, G. Manati, H. Ayu, W. Lereou, FU. Liogu, Hemsiyel, B. Mha, S. Ayu.

Identification TL 244–366 cm (8–12 ft), WT 363–454 kg (800–1 000 lb).
The only truly aquatic freshwater mammal in West Africa. The skin is naked, grey, and hairless, sometimes covered with green algae. The head is rounded with a large muzzle covered with whiskers, small round nostrils and very small eyes. The body is cylindrical, tapering into a tail ending in a dorso-ventrally flattened fluke. The fore limbs are short, modified into flippers, each with three srnall nails. The hind limbs are absent.

Distribution (a) Large permanent rivers of West Africa, e.g. Niger, Benue, Senegal; Lake Tchad, Volta

27

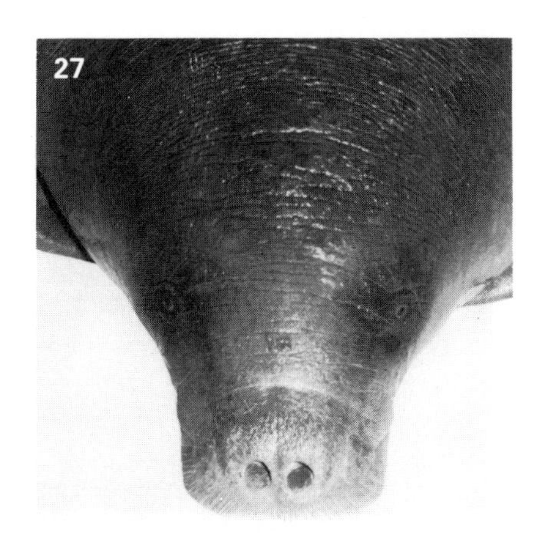
27

Lake and Kainji Lake. (b) Permanent rivers of Gabon, Central African Republic and the Congo. Two related species live in the rivers of the coastal south-east USA and north-east South America.

Habits Manatees are entirely aquatic and are unable to move on land. They live in large rivers and lakes where they feed on aquatic vegetation. They browse on the water weeds which are plucked with their large fleshy lips. Manatees swim slowly using their anterior flippers and an up and down movement of the tail. When resting they float just below the surface looking like a decaying log. The nostrils are on the top of the muzzle, and are closed when the animal submerges.

The gestation period is about five months; the single young is born in the water.

Status This defenceless animal has been over-hunted for its meat, and is now rare or very rare. It requires complete protection because of its rarity and its value in checking the growth of water weeds in large rivers and lakes.

Perissodactyla

The Perissodactyla are the 'odd-toed' ungulates since the weight of the body rests on one digit (the third), or on three digits (second, third and fourth) of each foot (cf Artiodactyla). The order contains several species which are not very similar in appearance, but they are all herbivores, have upper incisors, simple stomachs, and do not ruminate.

There are three families, two of them in Africa:

Equidae – zebras (Africa), horses (central Asia).

Tapiridae – tapirs (S. America and Malaya).

Rhinocerotidae – rhinoceros (Africa, India, Java, Sumatra).

Zebras are found in eastern and southern Africa, but they have never occurred in West Africa. Domestic horses and donkeys belong to the family Equidae.

Rhinocerotidae

The rhinoceroses are very large, heavily built animals with one or two horns, made of hardened hair, on the snout. The skull is elongated and held rather low, and the body is rounded with ridges of folded skin. The skin is hairless. The legs are short and thick with three or four digits on the forefoot and three digits on the hindfoot. Black rhinos browse on leaves and shoots using the upper lip to pull the leaves from trees. They are fond of bathing in water and mud so that the skin becomes covered by a layer of earth. The senses of smell and hearing are good but sight is poor. Despite their large size they can run very quickly. The horn is sometimes used in attack.

There are five living species. One of these, the black rhinoceros, is found in the savannas of Africa including a part of West Africa and another, the white rhinoceros, lives in a few localities in eastern and southern Africa.

28 Black rhinoceros

Diceros bicornis
F. Rhinocéros noir, G. Spitzmaul-Nashorn, H. Karkanda, FU. Ngorowa, Killifori.

Identification SH 140–178 cm (55–70 in), WT 909–1 363 kg (2 000–3 000 lb).
This is a very large heavily built animal with grey, hairless skin, small eyes and small rounded ears. Its head is relatively small, its muzzle narrow with two horns composed of hardened hair, the posterior horn smaller than the anterior one. The body is large and barrel-like, the back concave. The legs are short and solid and the feet are flattened with small nails. The tail is short.

Distribution (a) Limited Guinea savanna of northern Cameroon, formerly north-eastern Nigeria. (b) Limited savanna areas in Africa; montane forests in East Africa.

Habits Rhinos are usually solitary, although the calves remain with their mothers for some time. Rhinos prefer to remain in thick scrub, but in some places (e.g. Ngorongoro in Tanzania) they live on grassy plains. They browse on the leaves of bushes, plucking the leaves and shoots with the elongated upper lip. Rhinos make several vocal noises, one of which is a puffing noise often heard when rhinos are disturbed. Tickbirds, which remove ectoparasites from the skin, also help to warn the rhino of possible danger by their behaviour. Rhinos have well defined home ranges. They require water regularly and often roll in muddy swamps coating the skin with mud. Despite their size and weight, rhinos can run at up to about 50 km/h (30 mph).
The gestation period is 530–550 days; there is one young.

Status Very rare, and almost extinct in West Africa. Their numbers have declined rapidly in the rest of Africa due to hunting and poaching; they need total protection.

Artiodactyla

The Artiodactyla are a very varied order ranging in size from the royal antelope (2–3 kg or 4–7 lb) to the hippopotamus (up to 3 050 kg or 3 tons). They are the 'even-toed' ungulates since the weight of the body rests on two digits (third and fourth), or on four digits (second, third, fourth and fifth). In the most advanced artiodactyls, the legs are elongated and the digits reduced to the third and fourth ones only. In some, there is development of canine teeth to form tusks, and in others there are horns on the top of the head.

They are all herbivores. Several families have a special part of the stomach called the rumen which stores undigested food; this food is later passed into the mouth again, chewed, and finally swallowed. This process is called rumination. The stomach is very varied in structure: two, three or four chambered, and either ruminating or non-ruminating. Many species have bacteria and protozoa in the stomach and caecum which aid in the breakdown and digestion of grasses and leaves.

There are six families in West Africa; the basic differences are shown on p. 28.

Suidae

The pigs are medium-sized, and have solid-looking bodies and relatively slender legs. Hair covers most of the body, or is confined to a mane on the neck and back. The canine teeth are elongated, especially the upper ones, to form tusks which are used in defence when necessary. They are gregarious animals, often forming parties of up to 12 individuals. They are mostly vegetarians, feeding on roots, grasses, and bulbs which are rooted with the snout and perhaps with the tusks; the stomach is two-chambered and non-ruminating. Despite their rather squat appearance, they can run quickly.

There are three species in Africa, all of them occurring in West Africa.

29 Red river hog

Potamochoerus porcus
F. Potamochère. G. Flusschwein, Busch-Schwein, H. Jan Alhanzir, T. Kokote, Batafo, W. Mbam Khukh, FU. Kossewy, Gaduru, Y. Túrùkú, Túùkú, D. Leule, S. Birnga.

Identification SH 63–76 cm (25–30 in), HB 127 cm (50 in), T 38 cm (15 in), WT 54–81 kg (120–180 lb). A pig-like animal with an elongated face and a rather laterally flattened body, it is a bright rufous colour with long white hairs on the tips of the ears and on the dorsal mane. The forehead is sometimes a deeper rufous colour than the body. It has small warts on the sides of its face (cf warthog), and small short tusks which do not curve upwards along the sides of its face (cf warthog). The lower canines form the largest tusks (cf warthog, giant forest hog). There is a black tuft at the tip of the tail. The young have longitudinal yellow stripes.

Distribution (a) Forest areas from Senegal to Cameroon, extending northwards into gallery forests in the savanna. (b) Lowland forest and gallery forests in the savanna throughout Africa south of the Sahara.

Habits Red river hogs live in groups of 4–20 animals in thick forest where there is plenty of cover, and often

Family	*Digits forefoot**	*Digits hindfoot**	*Number of stomach chambers*	*Ruminates*	*Horns*	*Teeth characteristics*	*Species in W. Africa*
Hippopotamidae (hippos)	4	4	3	no	no	lower canines form tusks	2
Suidae (pigs)	2+2	2+2	2	no	no	upper and lower canines form tusks	3
Camelidae[++] (camels)	2 (elastic pads)	2	3	yes	no	incisors $\frac{1}{3}$, no tusks	1 (domestic)
Tragulidae (chevrotains)	2+2	2+2	3	yes	no	upper canines form tusks	1
Giraffidae (giraffes)	2	2	4	yes	yes (covered by hair)	no upper incisors, no tusks	1
Bovidae (cows and antelopes)	2+2	2+2	4	yes	yes, but absent in females of some species	no upper incisors, no tusks	35

* 2+2 = Two digits touch ground, two digits higher up on back of foot do not touch ground.

++ = Camels are indigenous to Asia; one species, the Arabian camel, is domestic in West Africa and therefore not described in detail.

close to streams or swamps. They wander over a large area looking for roots, berries, fruits and bulbs and they may also feed occasionally on frogs, reptiles and eggs. They can cause a lot of damage in farmland, foraging for yams, cassava and other local crops. They obtain the food by digging with the pointed nose and tusks. Red river hogs are active at night, and rest in the daytime. When running, they hold their long ears upright with the tips falling over. In the Niokolo-Koba Park, the activity and distribution of the red river hog overlaps with that of the warthog; in the rest of West Africa they seem to be ecologically separated.

The gestation period is about five months; there are 3–6 young.

Status Difficult to assess due to secrecy and nocturnal habits, but probably rare although locally common.

30 Warthog

Phacochoerus aethiopicus

F. Phacochère, G. Warzenschwein, H. Mugun Dawa, Gadu, T. Sanka, W. Mbam Alla, FU. Mbala Lode, Gaduru ladde, Y. Ìmàdò, D. Lefale, B. Lee, S. Birgnie.

Identification SH 76 cm (30 in), HB 152–177 cm (60–70 in), T 46 cm (18 in), WT 68–114 kg (150–250 lb). (Males usually much heavier than females.)
A pig-like animal with a large head and well developed tusks. The body is elongated, mostly without hair except for scattered bristles on the back and sides and a mane of coarse hair from the nape to the middle of the back. The skin is greyish, sometimes rufous due to mud. The head is large, with a flattened nose, large warts on the side of the head and in front of each eye. The tusks are often well developed forming a semicircular shape. The upper canines form the largest tusks. The legs are shortish and the tail is naked except for a tuft of coarse hairs at the tip. The eyes are dark and small; the ears small and rounded. The young are the same colour as adults, and without stripes (cf red river hog).

Distribution (a) Guinea and Sudan savanna from Senegal to Cameroon. (b) Savanna regions of Africa south of the Sahara.

Habits One of the commonest animals seen in the savanna regions, warthogs are usually seen singly or in family groups up to about 10. When moving about, they often trot quickly, with the tail held erect, often following each other in a line. Warthogs are mainly active during the early morning and evening, but may be seen at any time of the day. They feed on roots, bulbs, and shoots by digging into the ground with the muzzle. Sometimes they kneel down on the forefeet so that the mouth and tusks are closer to the ground. In Yankari Game Reserve, warthogs graze on short green grasses. Warthogs like to drink daily, and they are fond of mud baths. They will dig for water, and enlarge waterholes made by elephants. Their senses of hearing and smell are good, but eyesight is poor. They are not noisy animals, but they sometimes grunt when feeding and when alarmed.

The gestation period is about five months; there are 2–6 young, but the number of surviving young is rapidly reduced. In Nigeria, there appear to be two main breeding seasons each year.

Status Common in suitable localities.

31 Giant forest hog
Hylochoerus meinertzhageni
F. Hylochère, G. Riesenwaldschwein, T. Ebio, D. Le fing.

Identification SH 102 cm (40 in), HB 152–177 cm (60–70 in), T 38 cm (15 in), WT 136 kg (300 lb).
This is the largest member of the pig family in Africa. Heavily built with fairly long legs, it is covered with scattered long coarse black hair, forming a crest to the neck and along the back. The head has a wide elongated snout. The eyes are small and dark with a slit-like opening of the preorbital gland in front of them and a facial swelling, rather like a wart, below them. The ears are slightly pointed. The tusks are small and undeveloped, the upper canines forming the largest tusks. The tail has a tuft of black hair at the tip.

Distribution (a) Forests of Portuguese Guinea, Liberia, Ivory Coast, Ghana, Togo and Cameroon. Not recorded from forests of Sierra Leone, Nigeria or Guinea. (b) Forests of Congo basin, Ethiopia and Sudan, and highland forests of East Africa.

Habits Giant forest hogs live in parties of up to 12 animals in dense forest. They prefer to live in areas where there is water and marshes, and they like to wallow in water or mud. Forest hogs are mainly

nocturnal, and feed on grasses, fruits, roots and berries. Probably they do not dig for food like warthogs. In the Albert National Park in the Congo, parties of up to 20 forest hogs have been seen during the daytime.

The gestation period is about four months; there are 2–8 young.

Status Very rare. Little is known about this rather secretive species. It was one of the last large mammals to be described from Africa; the first specimen was obtained in Kenya in 1904, and the first West African specimens were seen in 1906 in Cameroon, and 1930 in the Ivory Coast.

Hippopotamidae

The hippos are large, dark-coloured, hairless mammals who spend much of their time in fresh water. The incisor and canine teeth are well developed, especially the lower canines which form tusks. Hippos feed on grasses on land at night but spend the daytime in the water. They can remain submerged for up to six minutes.

The weight of the body is supported by short thick legs and all four digits of each foot rest on the ground. The stomach is three-chambered, but non-ruminating.

There are two species in West Africa.

32 Hippopotamus

Hippopotamus amphibius
F. Hippopotame, G. Flusspferd, H. Dorina, T. Susono, W. Lebhere, FU. Ngabbu, D. Mali, B. Mali, S. Banga.

Identification SH 140–160 cm (55–63 in), HB 330–406 cm (130–160 in), T 38 cm (15 in), WT up to 3 050 kg (3 tons). An enormous semi-aquatic animal. Its body is very large and barrel-shaped, the skin is brownish-grey to pinkish, hairless except on the muzzle and inside the ears. The head is very large and the big mouth is capable of opening to more than 90 degrees. The nostrils are on top of the muzzle and the eyes protrude from the top of the head so that often only the muzzle, eyes and ears are visible when the animal is in water. The canine teeth are well developed, forming tusks which are not visible when the mouth is closed. The legs are short and thick-set. The tail is small with a slight tuft of hair at the end.

Distribution (a) Large permanent rivers of West Africa. Lake Tchad. (b) Permanent rivers and lakes of Africa south of the Sahara.

Habits Hippos are mainly aquatic and are good swimmers. They spend the daytime submerged in water, sometimes with the top of the head above the surface, sometimes completely submerged but coming up for air every four to six minutes. If disturbed, they remain submerged and walk along the bottom of the rivers, just poking their nostrils above the surface for air. At night, hippos leave the water and feed on grasses. They may travel several miles from water while searching for food. During the rainy season, hippos may move long distances to live in temporary pools and marshes, but they return to the deep parts of the rivers or lakes during the dry season. Hippos are gregarious animals, although the males may often fight and injure one another. Because of their enormous food requirements, hippos may be destructive to vegetation on land if they become too numerous. Occasionally, they damage crops in local farms. However, they increase the productivity of their aquatic habitat because they defaecate in the water, and stir up mud causing nutrients to be released.

Fish of the genus *Labeo* are sometimes seen swimming around the neck and back of hippos. These fish are reported to feed on algae on the skin of hippos, and on excreta.

The gestation period is about 230 days, and there is usually one young. Hippos reach reproductive maturity when four to five years of age.

Status Once very common, but now generally rare. They are still found in the Senegal, Niger, Benue and other permanent rivers of West Africa.

33 Pigmy hippopotamus
Choeropsis liberiensis
F. Hippopotame pygmee,
G. Zwergflusspferd, H. Wadan-Dorina, D. Melikununi.

Identification SH 79 cm (31 in), HB 152–177 cm (60–70 in), T 15 cm (6 in), WT 272 kg (600 lb). The pigmy hippo has roughly the same proportions as the hippo, but is considerably smaller and has a relatively smaller rounder head. The skin is naked, greyish to grey-black. The nostrils are large and circular. The eyes are on the side of the head (cf hippo). The teeth are not as well developed as the hippo's; there is only one pair of upper incisor teeth (cf hippo). The back is arched and the legs are rather short with digits spread out.

Distribution (a) Restricted swampy areas in high forest zone of Portuguese Guinea, Sierra Leone, Guinea, Liberia, Ivory Coast, S.W. Ghana and the delta region of the Niger river. (b) Not present elsewhere.

Habits Pigmy hippos are secretive animals, and little is known of their habits. They are less aquatic than the hippo, and when disturbed they disappear into dense vegetation by the sides of rivers and swamps. They make their own tracks and tunnels through the vegetation. Pigmy hippos are usually solitary animals. They may be territorial.

The gestation period is about 200 days; there is one young.

Status This is a very rare species, and is listed in the IUCN Red Book of endangered species. Special reserves are required for their protection and survival. They breed well in captivity (e.g. at Basle Zoo in Switzerland).

Tragulidae

The water chevrotains are a family of small bovids similar in general appearance to duikers, but without horns. The most noticeable characteristic is the upper canine teeth which are elongated to form small tusks. Water chevrotains are shy animals living in tropical forests, usually close to rivers or swamps. They are semi-aquatic. Chevrotains feed on fruits and aquatic plants; the stomach is three-chambered and ruminating.

There are four species: three in Asia and one in Africa.

34 Water chevrotain

Hyemoschus aquaticus
F. Chevrotain aquatique,
G. Zwergmoschustier, D. Giminan,

FU. Diaure ndiyam, Y. Isẹ̀ .

Identification SH 36 cm (14 in), HB 91–102 cm (36–40 in), WT 14–16 kg (30–35 lb).
A small antelope-like animal, rufous-brown with white markings. It has a small pointed head, rather a hunched back and hindquarters higher than its shoulders. The upper canine teeth are elongated to form small tusks which project downwards on either side of the lower jaw. The bright rufous coat has white spots in vertical lines on the back and flanks, with one or two horizontal lines from the shoulder across the flank to the rump, and two white patches on the throat separated by a brown patch. The underparts are white. The outer digits do not touch the ground.

Distribution (a) Forest zone of West Africa, near swamps and rivers, from Guinea to Cameroon. (b) Forest zone of Central African Republic, Gabon, Congo (B) to Kivu.

Habits Water chevrotains are semi-aquatic, and always found close to water or swampy areas in the forest. They are shy animals, and when disturbed, they plunge into water and often hide almost submerged. They are good swimmers. Water chevrotains are nocturnal and usually solitary. They are mainly herbivorous, but may eat some animal food.

There is usually one young.

Status Difficult to assess because of their habits, but probably rare.

Giraffidae

This family contains two species, the giraffe and the okapi. The horns are small and covered with skin and hair, and are unlike the horns of any other member of the order. The foot has two digits only. They feed on leaves, and because of their height, they are able to feed on vegetation which is too high to be obtained by other herbivores. The stomach is four-chambered and ruminating. The long legs have to be spread out when drinking so that the head can reach the surface of the water. The giraffe lives in the savanna, has a patterned coat of large brown patches and is very tall with a long neck and long legs; the okapi lives in dense forest, has a striped coat, is not so tall and has a shorter neck and legs.

There is only one species in West Africa.

35 Giraffe

Giraffa camelopardalis
F. Girafe, G. Giraffe, H. Rak̂umin Daji, Rak̂umin Dawa, W. Ndiamala, FU. Ndiamala, Tirewa, B. Tuminan, S. Bouré.

Identification SH 305–366 cm (10–12 ft), Top of head 427–549 cm (14–18 ft), T 76 cm (30 in), WT up to 1 180 kg (2 600 lb).
The body is covered with hairs forming a pattern of chestnut-brown patches on a yellow-beige background. The head is elongated with large fleshy lips, large dark eyes, a pair of short stumpy horns covered with skin and hair, and elongated ears. The neck is very long with a mane on the dorsal side. The body is relatively small and it has very long legs. The patterned effect extends down the legs as far as the knees. The underparts are pale, sometimes with few brown patches. The tail is short with a long tuft at the tip. The

West African subspecies, *peralta*, has a pattern of larger and more widely separated brown patches than other subspecies. There is considerable variation in the markings of different individuals, even within a single herd.

Distribution (a) Guinea, Sudan and Sahel savanna north of the Niger-Benue river system (except near the source of the Benue in central Cameroon where some giraffes occur south of the river in the Parc National de la Bénoué). (b) Savannas of most of Africa south of the Sahara.

Habits Giraffes are sociable animals, usually seen in herds of 4–20 animals. Their distinctive patterning often makes them difficult to see among the savanna trees where they feed using their prehensile lips to browse on the foliage. When walking, giraffes 'pace' with the legs on each side moving in synchrony, but when galloping the forelegs or the hindlegs move together resulting in a 'rocking horse' motion. The senses of smell and hearing are well developed. Giraffes sometimes make soft moans and snorts, but are usually silent. They drink irregularly and so are able to feed far away from a source of water. Sometimes a pair of male giraffes rub and hit their necks together; this 'necking' behaviour is correlated with sexual activity. Another interesting behaviour pattern, which is probably also of sexual significance, is 'urine testing' when the male samples the urine of the female. The long powerful legs are sometimes used for defence. The gestation period is about 450 days; there is one young.

Status Very rare in West Africa except in certain reserves, e.g. Parc National de Waza. Their numbers have been reduced by hunting during the last 40 years. In need of special protection.

Bovidae

This is the most important and abundant family of artiodactyls, with 35 species in West Africa.* They vary in size from the Derby eland (weighing 680 kg or 1 500 lbs) to the royal antelope (weighing 2–3 kg or 4–7 lb). In all species, the males have horns; in some species, the females have horns as well. The horns are always unbranched, have a bony core attached to the skull and an outer sheath of hardened keratin (the same substance as in nails and hoofs); there is great variety in the size and shape of the horns. The animal retains the same set of horns throughout life, even if a horn is partly broken or mis-shapen. The bovids are herbivorous and they either graze on grass, or browse on leaves and shoots. The stomach is four-chambered and rather complicated. The food is always ruminated.

The family can be sub-divided into several sub-families as follows:

Tragelaphinae – spiral-horned antelopes
Bovinae – buffalo
Cephalophinae – duikers
Reduncinae – waterbuck, kob, reedbuck
Hippotraginae – 'horse-like' antelopes
Alcelaphinae – hartebeest, korrigum

* Deer do not occur in Africa south of the Sahara. They belong to a separate family (Cervidae) and have antlers on the head which are branched and replaced each year. Deer are found in Europe, Asia, North and South America, and Africa north of the Sahara.

Neotraginae	– dwarf antelope, oribi
Antilopinae	– gazelles
Caprinae	– wild goats

Tragelaphinae

All members of this sub-family have spiral horns, either a single spiral as in bushbuck or many spirals as in the eland. There are horns in both sexes, except in bushbuck and sitatunga. The coat is usually patterned with white spots or stripes. In the genus *Tragelaphus* (bushbuck and sitatunga), the males are distinctly larger than the females.

The nine species are all African; four of them occur in West Africa.

36 Bushbuck

Tragelaphus scriptus
F. Guib harnaché, G. Schirrantilope, Buschbock, H. Mazo, Ganjar, T. Wansane, FU. Diaure, Njama chirga, Y. Ìgalà, D. Minan, B. Mina.

Identification SH 69–76 cm (27–30 in), HL 25 cm (10 in), WT 32–54 kg (70–120 lb).
A medium-sized antelope with short, dense hair. In colouring it is rufous to bright chestnut with a white spot on each cheek, two white patches on the throat, six to seven vertical white stripes on the shoulders, flanks and rump, two horizontal white lines on the lower part of each flank and a large white spot above each foot. It has large ears sticking out from the sides of the head. The horns in the male have a single spiral, extending slightly backwards and outwards and turning slightly forward at the tip. The West African bushbuck has more white markings than other subspecies, and the patterning of the white lines on the sides has resulted in the common name 'harnessed antelope'. Some individuals have very dark underparts.

Distribution (a) Forest zone and gallery forests throughout most of West Africa. Also relic forests in Guinea savanna (b) Forest and forest patches where there is fairly dense cover throughout Africa south of the Sahara.

Habits Bushbuck live close to water in country where there is plenty of cover. They are usually solitary or in pairs. When alarmed, bushbuck make a dog-like bark and run with a bouncing movement; the tail is held upright so that the white underside is clearly visible. They are mainly browsers but they also graze. They have definite home ranges, and spend most of the time within a fairly restricted area. The gestation period is about 220 days; there is one young.

Status Widespread and locally common. Bushbuck can withstand quite heavy hunting pressure, and their skins are frequently seen in local markets. Their shyness and adaptability has helped their survival.

37 Sitatunga

Tragelaphus spekei
F. Situtonga, guib d'eau, G. Sitatunga, Wasserkudu, H. Ragon Ruwa.

Identification SH 114 cm (45 in), HL 51–63 cm (20–25 in), WT 91–109 kg (200–240 lb).
Similar to the bushbuck, but larger with less distinctive white markings. The coat is rather long and shaggy. The males are chocolate-brown to grey-brown, the females rufous to

rufous-brown. There are white patches on the muzzle in front of the eyes, below each eye and ear, and on the throat, and indistinct white lines, often broken into spots, on the flanks. The colour and white markings are variable. The legs are relatively long and thin with elongated pointed hoofs. The large ears stick out from the sides of the head. The horns, in males only, extend backwards and outwards, with one spiral.

Distribution (a) Swamps and rivers in forest regions extending in some areas into savanna; also around Lake Tchad. Not recorded from forests of Guinea, Liberia, Ivory Coast or Ghana. (b) Forest and swamp areas in most of Africa south of the Sahara except much of East and southern Africa.

Habits Sitatunga are semi-aquatic antelopes, and are found only in wet swampy places. The long hoofs spread out and are adapted for walking on soft ground. Sitatunga wade about in swamps, and sometimes submerge completely except for the top of the head. They can swim when necessary. These adaptations for life in swamps have resulted in clumsy awkward movements on dry land. They are nocturnal and diurnal, and usually live singly or in pairs. There is one young.

Status These shy antelope are difficult to observe, and therefore there are few observations on their status and habits. Their distribution is limited by lack of suitable swampy habitats. Probably rare.

38 Derby eland
Taurotragus derbianus
F. Eland de Derby,
G. Riesenelenantilope, FU. Djinki,
B. Ulamissi.

Identification SH 152–175 cm (60–69 in), HL 75–102 cm (30–40 in), WT 680 kg (1 500 lb). The largest and most impressive antelope of West Africa, the Derby eland is very large and heavily built. In colour it is rufous to rufous-grey with 12–14 vertical white stripes on the sides of the body, and a black stripe along the middle of the back. The head has white upper and lower lips and a bright rufous patch on forehead. There is a small mane on the neck and a large dewlap. There are black markings around the knees and a black tip to the tail. Both sexes have large spiral horns, straight but diverging outwards slightly. The western subspecies is slightly larger than the eastern subspecies, is bright rufous, and has more vertical white side stripes.

Distribution There are two subspecies in West Africa:
the western subspecies, *gigas*, and the eastern subspecies *derbianus*.
(a) *gigas* is found in a small limited region where the borders of Mali, Senegal and Guinea join; derbianus in Guinea and Sahel savanna of northern Cameroon.

Extinct now in rest of Guinea savanna of West Africa.
(b) *derbianus* in savanna of Tchad, Central African Republic and Sudan.

Habits Derby eland are gregarious, usually living in herds of 10–20; although groups of 60 have been seen. They are grazers but also browse on leaves and shoots. During the day, they usually rest in the shade of trees; at night they feed and often move considerable distances. Eland are shy animals and usually run away if humans are near. They have good senses of smell and hearing. They are not aggressive animals, and the closely related eland of East and South Africa is easily domesticated.

The gestation period is nine months; there is one young.

Status Eastern subspecies: very rare except where protected, e.g. Parc National de la Bénoué, Parc National du Boubandjidah.
Western subspecies: very rare except where protected, e.g. Parc National de Niokolo-Koba. This subspecies is listed in the IUCN Red Book of endangered species.

Both subspecies are in need of absolute protection.

39 Bongo

Boocercus euryceros
F. Bongo, G. Bongo, T. Tromo, Tuminan, D. Sigui.

Identification SH 122–127 cm (48–50 in), HL 63 cm (25 in), WT 227 kg (500 lb).
A medium-large antelope, with very vivid and distinct markings, the

bongo is a bright chestnut colour with 12 or more narrow vertical white stripes on the sides. There are stiff hairs along the back, forming a black and white striped mane. The head is chestnut with a white chevron between the eyes, two white spots on each cheek, and a white patch on the throat. The underparts are black, the legs black with white markings, the tail white with a black tip. Both sexes have horns with one spiral, diverging backwards.

Distribution (a) High forest in Liberia, Sierra Leone, Ivory Coast and Cameroon. Scattered localities in Ghana and Togo. Not recorded from Nigeria. (b) Forests of the Congo basin, and montane forests of Kenya and possibly Uganda.

Habits Bongos are the largest of the forest antelopes. They are secretive, nocturnal and rarely seen. They move through the forest with the head held forward and the horns resting on the back of the neck. They browse on leaves.

Status They are probably very rare except where there are large areas of undisturbed forest. Their survival depends on the conservation of high forest.

Bovinae

This is a subfamily of the 'cow-like' artiodactyls, from which domestic cattle have originated. They are large heavily built mammals, with enormous wide horns covering the top of the head and spreading outwards. They are gregarious, often forming large herds. They graze, often on tall grasses, and they live near water since they need to drink daily.

There is only one species in Africa.

40 African buffalo

Syncerus caffer
F. Buffle d'Afrique, G. Afrikanischer Büffel, Kaffern Büffel, H. Bauna, T. Ekoo, W. Nagu Alla, FU. Eda, Mbana, Y. Ẹfọ̀n, D. Sigifing, B. Sigui, S. Hamdieye.

Identification SH 102–152 cm (40–60 in), HB 254 cm (100 in), T 76 cm (30 in), WT 318 kg (700 lb) in forest up to 818 kg (1 800 lb) in savanna, HL 65 cm (26 in), Greatest width of horns 65 cm or 26 in (in W. African savanna buffalo).
There are two subspecies of buffalo in Africa: *nanus* is the forest buffalo which is relatively small (SH 102–127 cm or 40–50 in), has small horns, and is rufous in colour; *caffer* is the typical savanna buffalo which is large (SH 127–152 cm or 50–60 in), has large wide-spreading horns, and is blackish in colour. The West African savanna buffalo is intermediate between *nanus* and *caffer*.

A large ox-like animal, it is heavily built with stout short legs and a large head. The muzzle is hairless, but the large ears have a fringe of hair on their lower edge. The large, barrel shaped body has a slight crest of hairs on the midline from the neck to the middle of the back. The legs are thickset and the tail has a tuft at the tip. Horns in both sexes are very large and heavy with a broad boss across the top of the head. They extend outwards, upwards and slightly backwards, curling inwards at tip.

The colour varies: light to dark tan in forest forms; black, light or dark tan, and all intermediate colours and patterning, in savanna forms.

Distribution (a) Guinea and Sudan savanna from Senegal to N. Cameroon. The true forest form, *nanus*, in forest regions. (b) Throughout suitable savannas and forests of Africa.

Habits Buffalo, or 'bushcows', are gregarious animals. In savanna areas they live in small groups or in herds of up to 100 individuals; in forest areas they live in groups of 5–10 animals. They are mainly grazers, but will also browse on shrubs. Buffalo are usually found close to water since they need to drink each day, and they enjoy wallowing in water or mud.

They feed in the morning and evening, and sometimes into the night as well; in the middle of the day they rest in the shade. Oxpecker birds, which feed on ectoparasites, are often seen clinging to the hair on the back and sides of buffalo, and egrets regularly feed on insects which have been disturbed by the movements of the buffalo. Sometimes buffalo make a low mooing sound, especially at night, but usually they are quiet. Their sight and hearing are poor, but their sense of smell is good.

The gestation period is about 11 months; there is one young.

Status Uncommon except in some protected areas where they may be abundant.

Cephalophinae

The duikers are a subfamily of small browsing antelopes, most of them 60 cm (24 in) or less at the shoulder. The back is usually arched, and the head is held close to the body. With the exception of female crowned duikers, both sexes have small horns. These are sometimes barely visible, especially in females, and most species have a small tuft of hair on top of the head between the horns. The suborbital gland is well developed with a slit-like opening below each eye. They live singly or in pairs. There are two genera of duikers: (1) the forest duikers, *Cephalophus* spp, which usually have a stripe or patch of colour on the neck or back contrasting with that of the body, and horns in both sexes, and (2) the savanna duiker, *Sylvicapra* which is the same colour all over, with horns only on the male.

Duikers occur only in Africa. There are 17 species, and 13 of them are found in West Africa (Table 1).

41 Striped duiker

*Cephalophus doriae**

F. Céphalophe zèbré, G. Zebra-Ducker, FU. Were.

Identification SH 41 cm (16 in), HL 4 cm ($1\frac{1}{2}$ in), WT 9–$13\frac{1}{2}$ kg (20–30 lb).
This is a small greyish-rufous duiker with about 12 vertical black stripes on the sides and rump. The underparts are lighter without stripes, the legs are rufous with black bands across the middle and the tail has white hairs.

Distribution (a) Forests of Sierra Leone, Liberia and western Ivory Coast. (b) ——.

Habits The striped duiker lives in the high forests in a very limited area of West Africa. There are few records or observations on this species because of its rarity.

Status Very rare.

* The correct scientific name for this species may be *Cephalophus zebra*.

Table 1

Duikers of West Africa

Species (in order of size)	*Shoulder height (cm)*	*Average Weight (kg)*	*Colour of head tuft (if present)*	*General colour*	*Colour of dorsal line (if present)*	*Other diagnostic features*	*Distribution in West Africa* F = *Forest,* S = *Savanna,* F-S = *Forest-savanna boundary*
C. rufilatus	35	11–14	black	rufous	blue-grey (wide)	—	F-S most of W. Africa
C. maxwelli	35–40	4–9	dark brown	blue-grey	absent	Forehead chocolate brown, white line above eye	F Senegal – W. Nigeria
C. monticola	35–30	4–9	dark brown	blue-grey	absent	as for *maxwelli*	F E. Nigeria – Cameroon
C. doriae	40	9–16	no tuft	oatmeal	absent	12 vertical black stripes on sides	F Liberia, Sierra Leone, west Ivory Coast
C. niger	50	9–16	rufous orange	black	absent	chestnut forehead	F Guinea – W. Nigeria

C. nigrifrons	50	18	black	deep rufous	absent	black stripe from nose to top of head	F	Cameroon east of Sanaga R.
C. leucogaster	50	18	rufous with black	golden brown	black, thin	white undersides	F	Cameroon
C. callipygus	55	16–20	rufous orange	rufous	black	—	F	Cameroon
C. ogilbyi	55	20	rufous black	rufous brown	black thin	—	F	Liberia, Ghana, E. Nigeria, Cameroon
C. dorsalis	55	20	no tuft	rufous brown	black	—	F	Guinea – Togo, E. Nigeria, Cameroon
S. grimmia	60	11–14	rufous black	grey	absent	—	S	Most of West Africa
C. jentinki	78	65	no tuft	pale grey	absent	Neck and throat black, whitish collar round shoulder	F	Sierra Leone, Liberia, Ivory Coast
C. sylvicultor	82	45–65	rufous orange	dark brown	yellow on rump	—	F	Senegal – Cameroon

42 Yellow-backed duiker

Cephalophus sylvicultor
F. Céphalophe à dos jaune, G. Riesenducker, Gelbrücken-Ducker, H. Muturun Kurmi, T. Okwaduo, Y. Gìdì-gìdì, D. Komisi.

Identification SH 84 cm (33 in), HL 13 cm (5 in), WT 45–64 kg (100–140 lb).
This is the largest species of the duiker subfamily. Heavily built and mainly dark brown in colour, it has grey on the sides of its head and well developed orange, rufous or black crest. The opening of the suborbital gland below the eye is very conspicuous. There is a characteristic yellow patch on the back becoming wider in the middle and on the rump. The size of this patch varies, but it may be quite small. Its hairs are erectile. In young animals the patch is white or whitish-yellow.

Distribution (a) Forest regions from Senegal to Cameroon; occasionally in forest relics in savanna north to about 8°N. (b) Forest areas in Africa extending into savanna as far as Sudan-Congo border, Uganda, western Kenya and Zambia.

Habits Yellow-backed duikers live only in forest regions, and they are seldom seen because they are very shy. They are mainly nocturnal, and they live singly or in small groups. They browse on forest leaves.

Status Very rare.

43 Jentink's duiker

Cephalophus jentinki
F. Céphalophe de Jentink, G. Jentinks Ducker.

Identification SH 79 cm (31 in), WT 64 kg (140 lb).
This large, heavily built duiker has a brownish-black head, neck and throat, a greyish muzzle and lips, and a light grey collar around the shoulders. The back and rump are grey-brown, and the inside of the legs light grey. The horns are large for a duiker, very slightly curved, extending backwards from the head.

Distribution (a) Scattered localities in forests of Sierra Leone, Liberia, and Ivory Coast. (b) ——.

Habits This duiker is only found in dense forest, and nothing is known about its habits.

Status Very rare. A few specimens were obtained at the end of the last century, and then none were recorded until about 1960. This species is listed in the IUCN Red Book of endangered species, and requires absolute protection.

44 Maxwell's duiker

*Cephalophus maxwelli**
F. Céphalophe de Maxwell, G. Blauducker, H. Gadan Kurmi, T. Otwe, Y. Ẹtu, D. Diafi, Forni.

Identification SH 36–41 cm (14–16 in), HB 66 cm (26 in), T 13 cm (5 in), HL 5 cm (2 in), WT $5\frac{1}{2}$–7 kg (12–15 lb).
This is a small slate-grey duiker, with a rounded back and cramped

* Some authorities consider *Cephalophus maxwelli* and *Cephalophus monticola* to be the same species. Only their geographical distribution seems to differ: *C. maxwelli* occurs in West Africa as far east as the Cross River, and *C. monticola* in the remaining forested areas of Africa; therefore the two species do not overlap. The supposed difference between the two species is based on the structure of a gland in the foot.

Common jackal

2 Side-striped jackal

3 Ruppell's fox

4 Pale fox

5 Fennec fox

6 Hunting dog

8 Libyan striped weasel

11 Clawless otter

13 Spotted hyaena

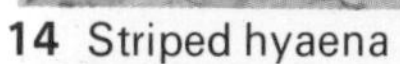

14 Striped hyaena

16 Sand cat

18 Caracal

19 African golden cat

20 Lion

21 Leopard

22 Cheetah

24 African elephant

25 Rock hyrax

26 Tree hyrax

28 Black rhinoceros

29 Red river hog

30 Warthog

32 Hippopotamus

33 Pigmy hippopotamus

34 Water chevrotain

35 Giraffe

36 Bushbuck

37 Sitatunga

40 African buffalo

41 Striped duiker

42 Yellow-backed duiker

43 Jentink's duiker

44 Maxwell's duiker

46 Black duiker

47 Black-fronted duiker

48 Red-flanked duiker

49 Peter's duiker

50 Ogilby's duiker

51 Bay duiker

52 White-bellied duiker

53 Crowned duiker

54 Waterbuck

56 Reedbuck

55 Kob

57 Mountain reedbuck

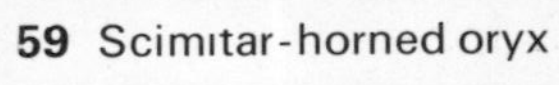

58 Roan antelope

59 Scimitar-horned oryx

60 Addax

61 Korrigum

62 Hartebeest

63 Klipspringer

66 Oribi

67 Dama gazelle

68 Dorcas gazelle

69 Red-fronted gazelle

70 Barbary sheep

appearance. The forehead is dark brown, with a dull white streak running above the eye from the base of each horn to the muzzle. There is a large slit-like opening of the suborbital gland below each eye surrounded by a bare patch. The head has a small black crest. The rest of the body is slate-grey to grey-brown. The tail is black, bushy and fringed by white hairs.

Distribution (a) Forests from Senegal to Nigeria as far east as the Cross river. (b) ——

Habits Maxwell's duikers are the commonest of the true forest duikers, but sometimes they are seen in patches of secondary forest. Observations in captivity show that they are territorial, and this is probably so in wild-living animals as well. The secretions from the suborbital gland are used by the male for marking his territory, and by both sexes during sexual behaviour. The gland has the same function in all duikers, but this is most easily observed in this species. Walking is rather jerky with the neck bent, the head close to the shoulders, and the tail flicking. When running, duikers hold their heads low. They feed on leaves, grains and fruits. The gestation period is about four months; the one young is normally born during the dry season.

Status Fairly common.

45 Blue duiker

Cephalophus monticola
See Note under *Cephalophus maxwelli*.

Identification Similar to *Cephalophus maxwelli* (not illustrated).

Distribution (a) Forest regions in Nigeria east of the Cross river, and in Cameroon. (b) Forested regions of Africa south of the Sahara, except west of the Cross river.

Habits Presumably similar to *C. maxwelli*.

Status Fairly common.

46 Black duiker

Cephalophus niger
F. Céphalophe noir, G. Schwarz Ducker, D. Tuba.

Identification SH 51 cm (20 in), HL 8 cm (3 in), WT 9–11 kg (20–25 lb).
A medium-sized duiker, with brownish-black hair, a streak of orange-rufous on the front of its head, an orange head tuft, and an orange patch between fore-legs. The underparts are paler than the rest of the body. The tail is brown-black and white underneath.

Distribution High forest from Guinea to western Nigeria.

Habits This species is found only in high forest. Little is known of its habits except that it feeds on leaves and grass and is mostly nocturnal.

Status Not uncommon in Liberia and western Ghana, but rare or very rare throughout the rest of its range.

47 Black-fronted duiker

Cephalophus nigrifrons
F. Céphalophe à front noir, G. Schwarzstirn-Ducker.

Identification SH 51 cm (20 in),

HL 5 cm (2 in), WT 18 kg (40 lb).
In general colour this duiker is deep rufous or chestnut, with underparts sometimes slightly paler. There is a black stripe from the nose to the top of the head, a black head tuft, and black on the lower part of legs. The tail is also black with a white tip.

Distribution (a) High forest of Cameroon east of the Sanaga river. (b) Forest of the Congo basin eastwards to Kivu.

Habits This species is found only in high forest, especially in marshy areas. Little is known of its habits.

Status Unknown.

48 Red-flanked duiker

Cephalophus rufilatus
F. Céphalophe à flancs roux, G. Rotflanken-Ducker, Blaurückenducker, H. Makurna, FU. Jabare (?), T. Akogyei, Asebee, Y. Èsúró, D. Konani.

Identification SH 36 cm (14 in), HB 66 cm (26 in), T 10 cm (4 in), HL 5 cm (2 in), WT 11–13½ kg (25–30 lb).
Orange-rufous in colour, with a broad blue-black line from the nose to the root of the tail, widening on the back but often indistinct. The legs are blue-black and there is a black head tuft. The ears are wide, very dark or black on the back. The tail has a black tip.

Distribution (a) Gallery forests and forest relics in Guinea and Sudan savanna up to about 13°N in Senegal, and northern margins of the forest zone throughout West Africa. (b) Similar habitats eastwards to East Africa.

Habits Red-flanked duikers live on the edges of forest, but not in the high forest nor in the open parts of the savanna. They hide in dense cover, but come out into the open savanna or openings in the forest to feed. They are most active in the early morning and at dusk, and are usually seen at these times. They are usually seen singly or in pairs, and they feed on leaves and grass. When disturbed they dash for cover with a bouncing run and the head held down. The alarm call is a shrill bark. There is one young, usually born in the dry season.

Status Locally common, but limited in distribution.

49 Peter's duiker

Cephalophus callipygus
F. Céphalophe de Peters, G. Schönsteiss-Rotducker.

Identification SH 56 cm (22 in), HB 89 cm (35 in), T 10 cm (4 in), HL 10 cm (4 in), WT 16–20 kg (35–45 lb).
The general colour is reddish-brown, slightly paler on underparts. The head tuft is orange-brown. A black line on the back begins between the shoulders, widens posteriorly, and ends on the underside of the tail.

Distribution (a) Forest zone of Cameroon, east of the Sanaga river. (b) Forests west of the Congo and Ubangi rivers as far east as Kivu.

Habits This species is found in high forest, and sometimes in secondary forest. Nothing is known about its habits in the wild.

Status Locally common.

50 Ogilby's duiker

Cephalophus ogilbyi
F. Céphalophe d'Ogilby,
G. Ogilby-Ducker.

Identification SH 56 cm (22 in), HL 10 cm (4 in), WT 20 kg (45 lb). In general colour this duiker is deep golden brown, the underparts paler than the back. There is a narrow black band from the middle of the back to the base of the tail. The head tuft is orange, the legs are dark and the tail has a grey tip.

Distribution (a) Forest zone of Liberia, Ghana, Nigeria east of the Niger river and Cameroon. (b) ——.

Habits No information.

Status Not known.

51 Bay duiker

Cephalophus dorsalis
F. Céphalophe à bande dorsale noir,
G. Schwarzrückenducker, T. Owio,
D. Turako-kuna.

Identification SH 56 cm (22 in), HL 5 cm (2 in), WT 20 kg (45 lb). The general colour of the bay duiker is ginger-brown. The upper lips and chin are white. A black stripe along the back begins at the top of the neck and ends at the base of the tail; in some specimens, the black stripe begins on the nose and extends to the tail with a break between the horns. The underparts are paler, with an obscure median brown line. The tail has a white tuft.

Distribution (a) There are two separate ranges: (i) forest zone of Guinea, Sierra Leone, Liberia, Ivory Coast, Ghana and Togo; (ii) forests of Nigeria east of the Cross river, and Cameroon. (b) Forests of Congo basin.

Habits Confined to forest. No other information available.

Status Distribution patchy, but locally common.

52 White-bellied duiker

Cephalophus leucogaster
F. Céphalophe à ventre blanc,
G. Weissbauchducker.

Identification SH 51 cm (20 in), HL 8 cm (3 in), WT 18 kg (40 lb). Though golden brown in general colour, the underparts, as the name indicates, are white. There is a faint stripe along the back beginning on the neck and ending at the base of the tail; this stripe becomes better defined posteriorly. The forehead is blackish-brown, with a rufous and black head tuft. The tail has a black and white tip.

Distribution (a) Forest of southern Cameroon. (b) Forest zone north of the Congo river as far east as the Kivu-Ituri area.

Habits Confined to forest. No other information available.

Status Not uncommon in suitable localities.

53 Crowned duiker

*Sylvicapra grimmia**
F. Céphalophe couronné, G.

* Other names for this species are Grimm's duiker, crested duiker, bush duiker and grey duiker.

Kronenducker, H. Gada, T. Sratwe, FU. Matakula, Y. Èkùlù, D. Mangarani, B. Cuntani.

Identification SH 46 cm (18 in), HL 10 cm (4 in), WT 11–13½ kg (25–30 lb).
Unlike the forest duikers this one does not have a curved back. Its general colour is grey, tending to rufous-grey on its anterior half. The hairs are flecked with black and white to give a peppered appearance. The head is rufous with a black stripe from the nose to the forehead. There is a prominent black head tuft. The ears are large, grey at the back. The animal's underparts are pale grey. Above the hoofs is a black ring. The tail is black above, white below. Compared with other duikers, it has relatively long legs and so appears to be lightly built. The female has no horns.

Distribution (a) Guinea savanna from Senegal to N. Cameroon. (b) Savanna of Africa south of the Sahara.

Habits Crowned duikers are the only real savanna duikers of West Africa. They live in open areas and in tall grass, and when disturbed they run away quickly, disappearing among the trees and grass. These duikers are seen singly or in pairs, and in the daytime they often feed or rest near the base of a tree. The colour of their hair makes it difficult to see crowned duikers in dead grass or burnt savanna. They browse on leaves and can go without water for long periods. The gestation period is about four months, and there is one young.

Status Locally common, especially in protected areas.

Reduncinae
This is a subfamily of medium or large antelopes. Only the males have horns; they are straight or curved, never spiral, and have well developed rings except at the tips. They are often seen grazing in grassy areas and fadamas close to water; they require regular water to drink, or succulent grasses.

There are four species in West Africa: waterbuck, kob, reedbuck and mountain reedbuck.

54 Waterbuck, Defassa waterbuck

Kobus defassa
F. Cobe defassa, Cobe onctueux, G. Defassa-Wasserbock, H. Gwambaza, Daddoka, T. Fusuo, FU. Bina, Ndumsa, Y. Òtòlò, D. Sense, B. Sine Sine.

Identification SH 117 cm (46 in), HB 203 cm (80 in), HL 64–76 cm (25–30 in), WT 181–204 kg (400–450 lb).
Shaggy but elegant-looking, this large antelope has fairly long, coarse hair which is brown or grey with white marks (often indistinct and varied) on nose, chin, throat, and above the eyes. The muzzle and the sides of the head are dark brown. The buttocks and the insides of hindlegs are white. The males alone have long, almost straight, horns, extending outwards and upwards from the skull and curving slightly forwards at the tip. They are heavily ringed except at the tip.

Distribution (a) Guinea and Sudan savanna from Senegal to N. Cameroon; may penetrate into forest edge. (b) Northern savannas of Africa southwards into Kenya (a similar species, the common waterbuck, is found from Tanzania south to South Africa; it may be conspecific with this species).

Habits Waterbuck live in wooded savanna or fadamas. They are fairly sedentary, remaining close to water since they drink regularly each day. Waterbuck may be seen singly or in large herds, these being comprised of females, or young males, or animals of both sexes. They are mainly grazers, but they sometimes browse when the grass is dry. Males are territorial, especially during the breeding season. Waterbuck have a musky smell due to secretions from glands in the skin. They are mostly silent, but will 'grunt' occasionally when disturbed.

Status Uncommon, except in reserved areas where they may be locally common.

55 Kob

Adenota kob
F. Cobe de Buffon, G. Schwarzfuss-Moorantilopen, H. Maraya, T. Frotee, W. Untu, FU. Kemba, Mbadda, Y. Egbin, D. Son, B. Son.

Identification SH 81 cm (32 in), HB 152 cm (60 in), T 28 cm (11 in), HL 30 cm (12 in), WT 68 kg (150 lb).
A medium-sized antelope with a short golden brown coat. There is a whitish patch around the eye to the base of the ear, and the underparts are white. The front edges of the forelegs are black. The males have 'S' shaped horns, going upwards, then bending sharply backwards and finally upwards again. They are heavily ringed at the base.

Distribution (a) Guinea and Sudan savanna from Senegal to North Cameroon. (b) Northern savanna, south to East Africa.

Habits Kob live in woodland savanna and open grassy plains near a permanent source of water. In the wet season, they move further from water since the fresh grass provides sufficient moisture; during the dry season, they remain close to rivers. They are usually found in herds of up to 40 individuals; these may be female herds, or young non-territorial male herds. Adult males are often solitary and territorial; they will fight for their territories, and males will 'herd' the females. These territories may include an earth mound or an old ant heap from where the male watches his territory. In the breeding season, there are herds exclusively of females and young. Kob sometimes leave their young in nursery groups of three or four (e.g. Borgu Game Reserve). During the daytime, kob do not retire into shady areas. They are usually quiet, although they sometimes give a short shrill whistle when disturbed. The gestation period is five-six months; there is one young.

Status Uncommon, except in reserved areas where they are locally abundant.

56 Reedbuck

Redunca redunca
F. Cobe de roseau, G. Riedbock, H. Kwanta-rafi, Kaji, T. Gyamframa, W. Mbill, FU. Padala, D. Kongoron, B. Kongoron.

Identification SH 71 cm (28 in), HL 20–25 cm (8–10 in), WT 36–45 kg (80–100 lb).

A medium-sized, graceful antelope with a beige, fawn or light rufous coat. The underparts are white. There are light rings round the eyes, and a bare, dark-coloured circular patch below each ear (as in oribi). The inside of the ears are black. The horns (males only) are straight, projecting slightly backwards with a hooked tip pointing forwards, and ringed at base. The tail is short, black on the upper surface, white on the lower surface. The subspecies *nigeriensis* (from Nigeria eastwards) has a black stripe on the front of the foreleg.

It is distinguished from oribi by its larger size and the shape of the horns, from crowned duiker by the presence of black spots below the ears, lack of head tuft, its slightly larger size and clear coloured coat without any peppered effect, and from female kob by the presence of the black spots below the ears and its smaller size.

Distribution (a) Guinea and Sudan savannas of West Africa where water and dense grass cover are available. (b) Northern savanna extending south to East Africa.

Habits Reedbuck are usually found in grasslands close to water, although they are sometimes seen in open woodlands. They are secretive and therefore difficult to see. They are seen singly or in small groups of up to four or five individuals. When disturbed, reedbuck run rapidly with the tail held erect so that the white underside is visible. They are grazers. There is one young.

Status Difficult to assess. Probably uncommon or rare.

57 Mountain Reedbuck

Redunca fulvorufula
F. Redunca de montagne, G. Bergriedbock, H. Kwanta-rafin Dutse, FU. Padala Hosere, Djabare Hosere.

Identification SH 74 cm (29 in), HB 107 cm (42 in), T 13 cm (5 in), HL 13 cm (5 in), WT 22–27 kg (50–60 lb).
Similar to reedbuck, but with smaller horns, a dark stripe on the nose, and thickset legs.

Distribution (a) Adamaoua mountains in North Cameroon (first recorded in 1962, and described as a new subspecies, *adamauae*). (b) Southern Sudan, Abyssinia, East and South Africa.

Habits Mountain reedbuck are found only in mountainous and hilly regions. They feed on grass but also browse. Very little is known about this species in West Africa, but they have been seen on the highest parts of the Adamaoua mountains. In other parts of their range they may be in groups of up to about 20 individuals.

Status Unknown, and very restricted in distribution. Probably uncommon or rare.

Hippotraginae

These are large antelopes with long straight, curved, or spiral horns in both sexes. Usually they have white and dark patterning on the face, and a mane on the neck. They live in savanna, semi-desert or desert habitats. Some species can exist with little or no water (addax, oryx), but others drink frequently (roan antelope).

There are six species in Africa, three of them in West Africa.

58 Roan antelope

Hippotragus equinus
F. Antilope rouanne, antilope cheval, Hippotrague, G. Pferdeantilope, H. Gwanki, T. Koo, W. Koba, FU. Koba, Y. Àgbá-ǹreré, Masia, Asanko, D. Koba, Dagbe, B. Dagwe.

Identification SH 140 cm (55 in), HL 74 cm (29 in), WT 227–272 kg (500–600 lb).
A large, horse-like antelope with the shoulders slightly higher than the hindquarters. The general colour is beige or dark tan. The head is black with white lips and chin and a white stripe above each eye. The ears are long, white inside, with a dark tuft at end. The horns, in both sexes, are thick and robust rising up from head in a sickle-shaped curve; heavily ringed. Young animals have small straight horns with only a slight curve. There is a chestnut-brown mane on the neck. The underparts are white. There is a blackish colouration on the legs and chest in some specimens. The tail has a long black tassel.

Distribution (a) Guinea and Sudan savanna from Senegal to Cameroon. (b) Savannas of Africa except south Africa and parts of eastern Africa.

Habits One of the most beautiful antelopes of the West African savannas, roan antelopes are usually seen in family groups and in herds of up to 30 or 40 individuals. Solitary males are not uncommon in Borgu Game Reserve. They tend to stay close to water. They are mainly grazers, moving to where fresh grass is available. In the dry season, they will browse on young green shoots; e.g. in Yankari Game Reserve roan antelope browse on mimosa bushes. The gestation period is about 10 months; there is one young.

Status Rare, but may be common in reserved areas.

59 Scimitar-horned oryx

Oryx dammah
F. Oryx algazelle, G. Säbelantilope, H. Mariri, FU. Ndubsa, B. Dangalan-Kule, S. Hundei.

Identification SH 122 cm (48 in), HL 102 cm (40 in), WT 204 kg (450 lb).
A large, heavily built antelope, pale sand to beige in general colour with a rufous-brown patch on the neck and chest. There are indications of pale rufous-brown on the back and flanks. The head has a brown stripe from nose to forehead. The scimitar-shaped horns are very long and ringed. The hoofs are large and the tail is long and tufted and black at the tip.

Distribution (a) Semi-desert from Mauritania to Niger. (b) Semi-desert eastwards to the Sudan.

Habits Oryx are adapted for living in arid environments. They are able to exist on very little water, and for long periods the water supply is solely from the food. They are very nomadic, and travel long distances in search of grasses and succulents. They migrate into areas where rain has fallen and grasses are sprouting. During the dry season, they migrate southwards to about 11 °N (in Tchad). During the middle of the day, oryx rest in the shade. They are gregarious, especially in the rainy season when herds of 12 have been recorded.

The gestation period is about nine months; one young is born in July–August.

Status Very rare. This species has been exterminated in most of its range due to indiscriminate hunting, often from motor vehicles. It is listed in the IUCN Red Book of endangered species, and is in need of absolute protection.

60 Addax

Addax nasomaculatus
F. Addax, G. Mendes-Antilope.

Identification SH 102 cm (40 in), HL 89 cm (35 in), WT 114 kg (250 lb).
A large, heavily built antelope, with rather short legs, and spiral horns. In general colour, it is pale sandy to sandy-grey, with white on the underparts and legs. The head has a white patch across the muzzle in front of the eyes, extending down the side of the face, and a brown patch on the forehead. Both sexes have long horns, with a single open spiral, ringed on the basal half.

Distribution (a) Desert and semi-desert from Mauritania to Niger. (b) Deserts of northern Africa.

Habits Addax are adapted for living in extremely arid environments. They are able to live without drinking at all, and can obtain all their water from their food. Like the oryx, they travel over large distances looking for grasses and succulents. Apparently addax are able to sense changes in the relative humidity of the air, so they can find areas where grass is sprouting or where rain has fallen. Sometimes they are seen in herds of up to 20 individuals. The gestation period is 10–11 months; there is one young.

Status Very rare throughout all of its range, and exterminated in many areas due to overhunting. It is listed in the IUCN Red Book of endangered species, and is in need of absolute protection.

Alcelaphinae

In this subfamily of medium-large antelopes, the shoulders are higher than the hindquarters so that the back slopes downwards. The horns are lyre-shaped or curved, and strongly ringed. There is great variety in the body colour and in the shape of the horns.

There are nine species in Africa, two of them in West Africa.

61 Korrigum

*Damaliscus korrigum**
F. Damalisque, G. Leierantilope, H. Dari, FU. Ndaura, S. Deritiré.

Identification SH 122–132 cm (48–52 in), HL 56 cm (22 in), WT 114–136 kg (250–300 lb).
A medium-large antelope, rather similar to the hartebeest. The shoulders are higher than the hindquarters so that the back slopes downwards. Its general colour is reddish-brown with bluish-grey patches on the forehead, muzzle, shoulders and rump. The underparts are also reddish-brown. Both

* Another common name for this species is Topi. Sometimes it is incorrectly called the Senegal hartebeest – but it is not a hartebeest and it occurs in many other countries besides Senegal (where it is now extinct).

sexes have horns rising from the head in a gentle curve, turning upwards and slightly inwards below the tip; and heavily ringed. It is distinguished from the hartebeest by its darker colour, bluish-grey patches, lack of high forehead and the shape of its horns.

Distribution (a) Guinea and Sudan savanna from Mali to Cameroon; distribution more northerly than for hartebeest but overlap in some areas. (b) Northern savanna eastwards to Sudan and East Africa.

Habits Korrigum are found in open wooded savanna and flood plains. They appear to be very similar to hartebeest in their habits and are usually seen in fairly large herds, but not in such numbers as the vast herds in East Africa. They are grazers. The gestation period is about seven–eight months.

Status Rare or uncommon although there are some large herds in protected areas.

62 Hartebeest, Western hartebeest

Alcelaphus buselaphus
F. Bubale, G. Kuhantilope, H. Kanki, T. Sonson, Burumu, FU. Kolongo, Lu'indu, Y. Ìrá kùnùgbá, D. Tankon, B. Tankon, S. Hira.

Identification SH 122–137 cm (48–54 in), HB 229 cm (90 in), HL 51–61 cm (20–24 in), WT 136 kg (300 lb).
A medium-large antelope with an elongated head and lyre-shaped horns. The shoulders are higher than the hindquarters so that the back slopes downwards. The general colour is beige to sandy-brown. The underparts are pale or white. The forehead is elongated due to the growth of a bony peduncle on the top of the head above the eyes. There are faint white lines between the eyes. The horns, in both sexes, rise upwards and slightly backwards in a 'U' shape, bending sharply backwards below the tip. They are thickly ringed, and those of the male are more robust and thicker than the female's and almost meet above centre of the head.

Distribution (a) Guinea and Sudan savanna from Senegal to Cameroon (western subspecies, *major*). (b) Northern savanna to Sudan, Abyssinia and parts of East Africa.

Habits Hartebeest live in grass plains and in open wooded savanna. They are seen in herds of up to 40 individuals of both sexes, and in male herds of up to 12 individuals (in Borgu Game Reserve). The males usually have territories in the breeding season. They are mainly grazers, and since they are not regular drinkers they can search for grasses far away from water. Often they are one of the first species to feed on new grass after burning. Sometimes they are rather shy and run away quickly, and at other times they stand and stare so that an observer can approach them easily. When disturbed, they make a short snort. The gestation period is about eight months; there is one young.

Status Uncommon, but locally common or abundant in reserved areas.

Neotraginae

A group of small antelopes with short pointed horns. Very varied in habits and habitats.

There are 18 species in Africa, four of them in West Africa.

63 Klipspringer

Oreotragus oreotragus
F. Oréotrague, G. Klippspringer, H. Gadar Dutse.

Identification SH 56 cm (22 in), HL 8–10 cm (3–4 in), WT 14–18 kg (30–40 lb).
A small compact antelope with thick, bristly hair that sticks out from the skin. Its general colour is yellowish-brown to greyish-brown, each hair speckled with yellow, brown and white. The head is small and rounded with small ears. The muzzle is brownish, the lips and chin are white. There is a black patch in front of each eye. There are small horns in both sexes, rising almost vertically from the skull and ringed at the base. There is no head tuft as in duikers. The underparts are white. The legs are thickset with 'rubbery' hoofs which have a black ring above them. The tail is very short.

Distribution (a) Rocky areas of the Jos plateau, Nigeria. (b) Limited rocky areas in southern and eastern Africa north to the Red Sea hills.

Habits Klipspringers are adapted for living on rocky surfaces in hills and mountains. They run up and down steep rocky slopes, and leap from rock to rock with abrupt jerky movements which are quite distinctive. Klipspringers are found in small groups. They are mainly browsers and do not drink regularly. The call is a shrill whistle. The gestation period is about seven months.

Status Very rare.

64 Royal antelope
Neotragus pygmaeus
F. Antilope royale, G.

Kleinstböckchen, T. Adowa, D. Sagbene.

Identification SH 25 cm (10 in), HB 38–51 cm (15–20 in), T 5–8 cm (2–3 in), HL $2\frac{1}{2}$ cm (1 in), WT 2–3 kg (4–7 lb).
This is the smallest of the African ungulates. Its fur is soft and sleek. The head, neck, back and sides are golden-brown becoming paler on the lower sides. The throat, chest, underparts, and inside of hindlegs are white, and there is a golden-brown 'collar' across the throat. On the front of each leg above the hoof is a white patch. The hind limbs are longer than the fore limbs, and are tucked under the body. This is an adaptation for high-speed running and jumping through dense undergrowth. The tail is fluffy, white except for some hairs on the upperside, and is constantly flicking. The males have small conical-shaped horns, without rings.

Distribution (a) Forests of Sierra Leone, Liberia, Ivory Coast, Ghana and Togo. Replaced by Bates' dwarf antelope on east side of the Dahomey Gap. (b) Occurs only in certain forests of West Africa, as above.

Habits Practically nothing is known about these antelopes in the wild because they are nocturnal, shy and rarely seen. They ruminate during the day. If disturbed, they either crouch motionless under cover, or move away at great speed. They bound over obstacles, or run smoothly with the head thrust forward, and at intervals they jump vertically high in the air.

Status These antelopes have a limited geographical distribution, and are rarely seen. Their status is unknown, but probably they are not uncommon in certain districts.

65 Bates' dwarf antelope

Neotragus batesi
F. Antilope de Bate, G. Bates-Böckchen.

Identification SH 30 cm (12 in), HL $2\frac{1}{2}$ cm (1 in), WT $5\frac{1}{2}$–7 kg (12–15 lb).
A very small antelope, with arched back and slender legs. Its general colour is deep chestnut, darker on the back. The underparts and throat are whitish. The tail is uniformly dark brown. The male has small straight horns, ringed at the base. There is no head tuft. It is distinguished from royal antelope by its slightly larger size, and different geographical distribution.

Distribution (a) Forest regions of Cameroon, and Nigeria east of the Niger river. (b) Lowland forest of northern part of Congo (K.) to the Uganda border.

Habits Very little is known about this species. They live only in lowland forest; and are rarely seen since they are shy and nocturnal. They live singly or in pairs.

Status Difficult to assess, but probably uncommon or rare.

66 Oribi

Ourebia ourebi
F. Ourébi, G. Bleichböckchen, H. Batsiya, T. Otwe, FU. Dyabare, Jabare, Y. Asiaro, Ogoro, D. Ngoloni, B. Nkolon.

Identification SH 51–66 cm (20–26 in), HL 10 cm (4 in), WT 14–18 kg (30–40 lb).
A small delicate-looking antelope with slender neck and long legs. In colour it is greyish-orange to pale chestnut, with white underparts. The forehead is deep rufous, and there is a black circular patch of bare skin below each ear (as in reedbuck). The ears are large and stick out from the side of the head. The tail is black or dark brown, white on the underside, and conspicuous when running. The horns in the male are straight, rather far forward on the head.

Distribution (a) Guinea and Sudan savanna from Senegal to Cameroon. (b) Northern savannas eastwards to southern Sudan and Uganda.

Habits Oribi live in open woodland and grassy areas, often where the vegetation is rather sparse. They often stand motionless under trees, but if disturbed they run away quickly with a long gait, leaping up and down in the grass. Oribi are usually seen singly or in pairs, but occasionally as groups of three or four (e.g. in Borgu Game Reserve). They are grazers, and usually remain close to water. When disturbed, oribi make a shrill whistle. The gestation period is about seven months; there is one young.

Status Uncommon to rare, although common in some protected areas.

Antilopinae
The gazelles are a group of small to medium-sized slender antelopes with long necks and long thin legs. They are fast runners. Horns occur in both sexes, and are lyre-shaped and thickly ringed (except in some species not occurring in West Africa). Some species can go without water for long periods of time, and are adapted for desert environments.

There are three species in West Africa.

67 Dama gazelle, Addra gazelle

Gazella dama
F. Gazelle dama, G. Dama-Gazelle, H. Farin-gindi, FU. Lelwa, S. Sanaï.

Identification SH 91–102 cm (36–40 in), HL 30 cm (12 in), WT 73 kg (160 lb).
A medium-sized antelope. In colouring, the neck, shoulders and back are chestnut. The head, legs, hindquarters and underparts are white except for a wide chestnut band across the flanks. There is a white patch on the front of the neck and a chestnut stripe on the front of the forelegs. The horns in both sexes are bent sharply backwards, so that they lie parallel to the back, turning upwards at the tip; they are heavily ringed.

There is great variation in the amount of chestnut colouration. The eastern specimens have more white and less chestnut than the western ones.

Distribution (a) Semi-desert and Sahel savanna from Mali and Mauritania to Niger. (b) Southern Morocco and semi-deserts south of the Sahara eastwards to the Nile valley.

Habits Dama gazelles are adapted to a semi-desert environment, and can go without water for long periods of time but not as long as the dorcas gazelle. They browse on bushes and shrubs, often standing on their hind legs to obtain the shoots higher up. These gazelles are found singly or in herds of up to 15. Previously they were commoner than at the present time: herds of several hundred dama gazelles were seen moving southwards during the dry season.

Status Numbers have been reduced considerably in recent years, and they are now rare.

68 Dorcas gazelle

Gazella dorcas
F. Gazelle dorcade, G. Dorcas-Gazelle, H. Farin barewa, FU. Lelwa ndanewa.

Identification SH 53–61 cm (21–24 in), HL 25 cm (10 in), WT 23 kg (50 lb).
A small-medium gazelle. Its general colour is sandy-brown to pale reddish-brown with a faint wide reddish-brown stripe on the flanks. The underparts and rump are white. There is a deep reddish-brown stripe from nose to forehead, and from the mouth to the inner corner of each eye. The tail is black. The horns in both sexes rise upwards, then slightly backwards, and finally upwards at the tip; they are heavily ringed.

Distribution (a) Desert and semi-desert south to about 13°N from Mauritania to Niger. (b) Deserts and semi-deserts of northern Africa, Arabia and parts of the Middle East.

Habits Dorcas gazelles are able to live in severe desert environments due to their ability to survive without drinking for long periods of time. They are very nomadic, wandering over large areas looking for grass and shrubs. They will drink when water is available. They are gregarious and may be seen in herds of up to 20 individuals. The gestation period is about three months; there is usually one young.

Status Uncommon to rare. Their numbers have been reduced in recent years, but this species has survived better than the other desert gazelles.

69 Red-fronted gazelle

Gazella rufifrons
F. Gazelle à front roux, Gazelle corine, G. Rotstirn-Gazelle, H. Barewa, Jan Barewa, W. Kevele, FU. Lelwa, S. Dieri:

Identification SH 66 cm (26 in), HL 25 cm (10 in), WT 27 kg (60 lb).
A medium-sized gazelle with a long neck, distinguished from the dorcas gazelle by its straighter horns and the black stripe on its flanks. Reddish brown, with white underparts and rump, it has a wide black stripe on the flanks above the white of the underparts. There is a small white eye ring, with a faint white line

running from the eye down to the mouth. The horns curve slightly backwards and then forwards, and are heavily ringed. The tail is black.

Distribution (a) Sudan and Sahel savanna, and semi-desert, from Senegal to Niger. (b) Similar habitat eastwards to the Sudan.

Habits This is the least desert-adapted of the gazelles. They prefer open areas to thick woodland, even though their distribution is further south than that of the other gazelles in West Africa. They live in small groups of 5–10 individuals. They are grazers and browsers. When disturbed they utter a wheezy snort several times in quick succession. The gestation period is about five months.

Status Rare, since they have been exterminated in much of their range.

Caprinae

The sheep and goat subfamily are medium-sized artiodactyls with large curved horns in both sexes. The African species live in dry rocky areas in desert or semi-desert environments. The hair is mostly a uniform brown or sandy colour with tufts of long hair on the chin or neck.

There are two African species: the ibex in the Red Sea Hills and parts of Ethiopia, and the Barbary sheep in scattered localities in the Sahara desert and semi-desert south of the Sahara.

70 Barbary sheep

Ammotragus lervia

F. Moufflon à manchettes,
G. Mähnenschaf, H. Ragon duci,
FU. Edda.

Identification SH 102 cm (40 in), HL 71–76 cm (28–30 in), WT 104 kg (230 lb). Similar in form and proportions to a goat, the Barbary sheep is sandy-brown with paler underparts. It has a small mane on the dorsal part of the neck and between the shoulders. It has long flowing hair on the throat, chest and forelegs, and on the tail. The horns, in both sexes, are very stout and circular in section, curving upwards, backwards and then downwards to form a semicircle over the neck; there are partial rings on the front edges of the horns, often worn in adults.

Distribution (a) Rocky areas in the desert and semi-desert region from Mali to Niger. (b) Rocky areas in desert and semi-desert eastwards to Sudan, Egypt, Eritrea and Arabia.

Habits Barbary sheep live in rocky areas where their colour makes them very difficult to see amongst the rocks and boulders. Usually they live in small groups of a male with several females and young, but occasionally they form larger herds. They are very agile and can jump from rock to rock. During the daytime they rest in the shade, but at night they graze and browse among the rocks or in sandy areas at the base of the rocks. Barbary sheep can survive without water for long periods, but will drink freely when water is available.

Status It is unlikely that Barbary sheep have ever been common because of their limited habitat in the desert environment. Their numbers have been reduced by overhunting and they are now uncommon or rare.

Conservation of Large Mammals

It will be obvious to most readers that many of the species described in the preceding pages are now rare, and their numbers and ranges have been reduced greatly in the last fifty years or so. The few studies on why there has been such a reduction suggest that overhunting has been the main cause. In many parts of West Africa, local hunters with home-made guns shoot animals all the year round with no regard for the sex, age or numbers that they shoot. Game laws are disregarded, and there is no-one to supervise and check that the laws are enforced. Many years of this indiscriminate destruction have resulted in the present scarcity of so many species. Besides the local hunters, there are a few licensed hunters with proper firearms who shoot a certain number of animals each year, but this is regulated and has a small effect on the animal populations as a whole. There are considerable differences in the amount of 'local' and 'professional' hunting and the efficiency of enforcing the wildlife laws in different West African countries. But in general it is true to say that uncontrolled local hunting with guns, nets, snares, and the use of fire (all of which are illegal) is the main reason for the present low numbers of large mammals in West Africa.

Many species of mammals, large and small, are able to increase their numbers quite rapidly if unmolested and living in an area which contains all the requirements for their existence and reproduction. In West Africa there are several national parks and game reserves where the natural vegetation and animals are undisturbed. Here many of the species described in this book can be seen easily; they do not run away when they see a car because they have learnt that it does not harm them. After visiting a reserve and seeing herds of buffalo, antelopes, elephants and many other animals in their natural environment, the visitor realises how few animals, if any, are seen outside the reserve. This is a very clear indication of the value of conservation, and how the country outside reserves has been altered and disturbed as a result of man's activities. The pages that follow list the major conservation areas in West Africa and the mammals which live there. Some of these reserves are well developed and carefully looked after; others suffer from a lack of funds and qualified personnel who know how to manage and look after a conservation area.

These reserves and parks show that large and flourishing wild animal populations can live in West Africa. Quite a lot of people have suggested that the West African environment, especially when compared with East Africa, is not capable of supporting large populations of large mammals. This is not true; the present condition, varying from country to country, is due to overhunting and the lack of foresight and determination to do anything about conservation in the past. French-speaking countries of West Africa have the best reserves; some of these were established in the 1930s and many years of conservation have resulted in large populations which are in balance with their environment.

As more and more of West Africa is altered by man and the natural vegetation is destroyed, it is even more necessary to have areas which are protected from unwise exploitation, and to decide what role the large mammals should play in the life and economy of the nation. I think the most important values of wild mammals are: enjoyment, educational, and economic.

Enjoyment For some people the enjoyment of watching and observing wild mammals is a good enough justification for conserving them. The majority of people, of all nationalities, enjoy their first visit to a reserve and are impressed when they see large numbers of wild mammals grazing contentedly a few yards away. For West Africans, the idea of *looking* at animals is a new one, but it is something that is enjoyed immensely at zoos, and this enjoyment and appreciation should be encouraged in parks and reserves as well. In many countries of the world, scenic areas and their animals are looked after with as much care as old historic buildings, sculpture, and other arts and crafts since they are all part of the national heritage. Why not in West Africa as well?

Some people enjoy hunting, but this is only possible if there are animals to hunt; and this in turn depends on a sensible policy to ensure that there is always a healthy population of wild mammals. If there is careful management, a certain percentage of the population can be removed each year by hunting without any adverse effects on the species as a whole. This careful utilisation can continue indefinitely, but only if it is regulated scientifically.

Education The value of wild animals is linked with their relationships with the plants, soil, water, and other animals living in the same area; the study of these relationships is known as 'ecology'. Knowledge about the ecology of wild mammals helps us to understand how they live and what effects they have on their environment. The value of ecological studies is most obvious in places where simple ecological principles have been disregarded and have led to erosion, loss of soil fertility, a reduction in the numbers of beneficial animals, and a degradation of the environment. Ecology is now an important part of biology syllabuses, and much of the new 'O' and 'A' level courses in West African schools is ecological in outlook. The best way to learn ecology is to go out and see the inter-relationships of animals and plants in their natural environments. Reserved areas, with their full spectrum of animals and plants, are the best 'outdoor laboratories' for ecological education. Every country should have many reserved areas – generally as large as possible – in each of the vegetation zones, where the adaptations of animals and plants to their environment can be studied.

One of the best illustrations of how well large mammals are adapted is to compare them with domestic cattle living under the same conditions. The plump healthy antelopes and buffalo are a striking contrast to the thin bony cattle that roam the savannas of West Africa. This is because wild mammals have evolved over a period of millions of years to live and survive in savanna; domestic cattle have not. Although cattle are important to man, wild mammals are ecologically much better at living in natural savanna and could be of greater value to man than at present.

Economic Since no individual owns the wild animals of a country, no-one really cares about their welfare. Consequently it is difficult to say if they have any economic value. But in most of West Africa, wild mammals are an important protein food. Estimates of the amount of 'bush meat' consumed in relation to total protein consumption varies from 20 – 85 per cent depending on the locality. In some places, then, wild animals are more important than domestic ones. A cow, a sheep, or a goat is worth so much money; similarly every wild mammal is worth so much. One study in 1965–66 suggests that the value of bush meat consumed each year in southern Nigeria is about £10 million; and so for the whole of Nigeria the total value is probably double this amount. Similarly in Ghana, a conservative estimate for the value of bush meat per year is over £3 million.

The irony is that the tremendous value of this important resource is not realised and hardly any funds are allocated for research, development and wise utilisation. It is high time that wild mammals were recognised as a natural resource as valuable as forests, minerals, agricultural products, and domestic animals. In areas not used for agriculture, the existing wildlife if protected for a number of years will increase to the point where a certain percentage of the animals can be cropped each year to provide a continuous supply of protein. This is an acceptable form of land-use in other parts of Africa.

There are other economic values besides protein production which are related to education and enjoyment. Tourism can provide employment for a lot of people, and there are many visitors who will pay large sums to come to Africa because of the scenery, climate and wildlife. Although West Africa is unlikely to become as popular as East Africa, some idea of the tourist value of wild animals and scenery is shown by the £16·7 million earned by Kenya in 1969 from tourism. In French-speaking West Africa, especially in Senegal and Cameroon, the tourist value of wild animals is already well established. Similarly internal tourism could be developed so that the West African, especially the younger generation, can benefit from the wild animals of his country, and indirectly this will help in the conservation of the animals themselves.

There are also many side products derived from wild mammals – skins, trophies, leather work and the export of wild animals; all these are additional economic values.

These examples show that wild mammals are of much greater value to a country than is realised. If nothing is done to ensure the survival of these animals and the environment which is essential for their survival, they will disappear in time and an important natural resource will be lost for ever. Alternatively an enlightened conservation programme will mean enjoyment, education, and economic advantage to everyone. But time is running out because of years of misuse and over exploitation. Conservation of wild mammals is needed *now* if they are to become the important resource that they deserve to be in the future. They are also part of the national heritage – a living example of the wonderful variety of animal life that has evolved in the forests, savannas and semi-deserts of West Africa.

National Parks and Game Reserves of West Africa

The principal conservation areas are listed alphabetically for each country. The information recorded for each park or reserve gives some idea of the vegetation, mammals, special features and facilities for visitors. The lists for some areas are accurate and complete, but for other areas they are incomplete and undoubtedly many other species occur as well. Unfortunately, no information is available for five of the sixteen West African countries – Portuguese Guinea, Guinea, Liberia, Sierra Leone, and Mauritania. The small mammals (mainly rodents, bats, and insectivores) are omitted, and similarly birds and reptiles which are usually an obvious and important part of the fauna are not included.

The abundance of mammals is indicated by:
a = abundant
c = common
r = rare
The visitor is almost certain to see 'abundant' species, and will probably see the 'common' ones. The 'rare' species are not often seen although they certainly exist.

Besides the species described in this book, there are other mammals which may be seen and which are described in *Small Mammals of West Africa*. The following list records the common and scientific names of these species, and the page where the description and information is found.

		Information in *Small Mammals of West Africa* page
Primates		
Senegal galago	*Galago senegalensis*	18
Potto	*Perodicticus potto*	16
Baboon	*Papio anubis*	29
Green monkey	*Cercopithecus aethiops*	23
Mona monkey	*Cercopithecus mona*	24
Spot-nosed monkey	*Cercopithecus nictitans*	26
Diana monkey	*Cercopithecus diana*	25
Patas monkey	*Erythrocebus patas*	27
Black colobus	*Colobus polykomos*	22
Red colobus	*Colobus badius*	20
Chimpanzee	*Pan troglodytes*	30
Pholidota		
Long-tailed pangolin	*Manis longicaudata*	—
White-bellied pangolin	*Manis tricuspis*	33
Giant pangolin	*Manis gigantea*	34
Lagomorpha		
African hare	*Lepus capensis*	36
Rodentia		
Flying squirrel	*Anomalurus spp.*	38
Sun squirrel	*Heliosciurus spp.*	42
Ground squirrel	*Xerus erythropus*	45
Giant rat	*Cricetomys gambianus*	55
Cane rat	*Thryonomys swinderianus*	59
Brush-tailed porcupine	*Atherurus africanus*	58
Crested porcupine	*Hystrix cristata*	57
Carnivora (Family Viverridae)		
Civet	*Viverra civetta*	62
Palm civet	*Nandinia binotata*	64
Forest genet	*Genetta maculata*	63
Blotched genet	*Genetta tigrina*	63
Pseudogenet	*Pseudogenetta villiersi*	—
Slender mongoose	*Herpestes sanguineus*	66
Egyptian mongoose	*Herpestes ichneumon*	68
Gambian mongoose	*Mongos gambianus*	65
White-tailed mongoose	*Ichneumia albicauda*	67
Marsh mongoose	*Atilax paludinosus*	67
Kusimanse mongoose	*Crossarchus obscurus*	66

Cameroon*

Parc National de la Bénoué

Locality Northern Cameroon, along the western bank of the Benue river.

Area 1 800 sq km (484 sq mi).

Formation 1936.

Vegetation Sudan savanna with *Isoberlinia, Monotes, Burkea, Daniellia, Uapaca, Combretum* spp, *Terminalia* spp, *Anogeissus, Kigelia, Khaya, Diospyros, Afzelia, Lophira, Ficus* spp.

Special features Benue river.

Access By road from Garoua 150 km (93 mi) and Ngaoundéré 150 km (93 mi).

Facilities Full accommodation, with running water and electricity, for 24 visitors in double rooms. Guides but no vehicles. Radio telephone.

Tracks in reserve 120 km (74 mi).

Open 25 December – 15 June.

Annual rainfall 1 200 mm (48 in).

Visitors 1969 450.

Principal mammal species:

Baboon a
Patas monkey a
Green monkey c
Black colobus r
Common jackal c
Hunting dog r
Spotted hyaena c
Serval c
Caracal r
Lion c
Leopard c
Cheetah r
Elephant c
Black rhinoceros r
Warthog c
Red river hog r
Hippopotamus a
Giraffe r
Buffalo a
Bushbuck c
Derby eland a
Red-flanked duiker c
Crowned duiker a
Kob a
Waterbuck a
Reedbuck a
Roan antelope a
Korrigum r
Hartebeest a
Oribi c

*Three other reserves and three sanctuaries (one each for gorillas, buffalo, and elephants) are planned for West Cameroon.

Parc National du Boubandjidah

Locality Northern Cameroon, south-east of Garoua and about 80 km (50 mi) east of the Parc National de la Bénoué.

Area 2 200 sq km (836 sq mi).

Formation 1947.

Principal vegetation Sudan savanna with *Isoberlinia, Burkea; Uapaca, Combretum* spp, *Terminalia* spp. *Kigelia, Khaya, Diospyros, Lophira, Afzelia, Ficus* spp, and borassus palms.

Special features Many salt licks.

Access By road from Garoua 280 km (174 mi) or Ngaoundéré 290 km (179 mi).

Facilities Full accommodation with 32 beds in double rooms; running water and electricity. Guides, but no vehicles. Radio telephone.

Tracks 300 km (186 mi).

Open 25 December – 15 June.

Annual rainfall 1 200 mm (48 in).

Visitors 1969 250.

Principal mammal species:

Baboon a
Patas monkey a
Green monkey c
Black colobus r
Common jackal c
Hunting dog r
Spotted hyaena c
Serval c
Caracal c
Lion c
Leopard c
Cheetah r
Elephant c
Rock hyrax r
Black rhinoceros r
Warthog c
Hippopotamus r
Giraffe r
Bushbuck c
Derby eland a
Buffalo a
Red-flanked duiker c
Crowned duiker a
Kob r
Waterbuck a
Reedbuck c
Roan antelope a
Korrigum r
Hartebeest a
Oribi a

Parc National de Waza

Locality Northern Cameroon, 250 km (150 mi) south of Lake Chad.

Area 1 700 sq km (666 sq mi).

Formation 1936.

Vegetation Mostly Sahel savanna with *Acacia* spp, *Balanites, Anogeissus, Tamarindus, Mitragyna, Ficus* spp, and Dum palms.

Special features Game-viewing hides, and four tree hides by marshes and lakes.

Access By road: 135 km (85 mi) from Fort Lamy; 120 km (74 mi) from Garoua.

Facilities Full accommodation for 90 visitors in double bedrooms at l'Hotel de Waza; running water and electricity. Game guides, but no game-viewing vehicles. Air strip. Radio telephone.

Tracks 450 km (279 mi).

Open 15 December – 1 June.

Annual rainfall 700 mm (28 in).

Visitors 1969 5 000.

Principal mammal species:

Baboon c
Patas monkey c
Green monkey r
Mongoose spp c
Zorilla r
Ratel r
Common jackal r
Spotted hyaena c
Striped hyaena c
Serval c
Caracal r
Lion c
Leopard r
Cheetah r
Elephant a
Rock hyrax r
Warthog a
Giraffe a
Bushbuck r
Crowned duiker c
Kob a
Waterbuck c
Reedbuck a
Roan antelope a
Korrigum a
Red-fronted gazelle c

Kimbe River Game Reserve

Locality Wum Division of West Cameroon. 42 km (26 mi) south of Nkambé. 1 500 m (5 000 ft) above sea level.

Area 40 sq km (19 sq mi).

Formation 1963.

Vegetation Montane grassland with riverine forest along water courses.

Special features Mekumba natural stone bridge, Kumte pool (good for fishing).

Access By road: Bamenda station – Banso – Nkambé; or Bamenda station – Wum.

Facilities Resthouse accommodation with beds, tents and cooking facilities. No catering. Guides.

Tracks in reserve About 6½ km (4 mi).

Open All year.

Annual rainfall No record; May–September.

Visitors 1969 About 330.

Principal mammal species:
Baboon a
Spot-nosed monkey r
Green monkey c
Mona monkey c
Ground squirrel r
Cane rat r
Civet r
Rock hyrax r
Buffalo c
Bushbuck r
Crowned duiker r
Kob a
Waterbuck a

Mbi Crater Game Reserve

Locality Wum Division of West Cameroon, north of Bamenda 1 800 m (6 000 ft).

Area About 1·3 sq km (½ sq mi).

Formation 1963.

Vegetation Montane grassland and marshes in old volcanic crater. Fringes of crater with forest of *Piptandeniastrum*, *Schefflera*, *Ficus*, *Paulyscias* and crotons.

Special features Old volcanic crater, caves.

Access By road Bamenda–Sabga 30 km (19 mi), and then by footpath 12 km (7 mi).

Facilities Native huts are available near the reserve.

Tracks in reserve None.

Open All year.

Annual rainfall No record; May–September.

Visitors 1969 About 30.

Principal mammal species:

Baboon	a	Rock hyrax	c
Green monkey	c	Bushbuck	c
Cane rat	r	Sitatunga	r
Civet	r	Duikers	c

Dahomey, Upper Volta, and Niger

Parc National de la Boucle de la Pendjari

Locality Northern Dahomey, on border with Upper Volta.

Area 2 750 sq km (1 045 sq mi).

Vegetation Northern guinea savanna with *Khaya, Pterocarpus, Parkia, Tamarindus, Daniellia, Buterospermum, Anogeissus*, etc.; extensive grass plains, gallery forest near river.

Facilities Full accommodation at Porga campement at entrance to park; running water and electricity. Hotel de la Pendjari in park; running water, electricity, air-conditioning, swimming pool. Airstrip at Natitingou (South of Park) and at Porga. Vehicles at Porga and Hotel de la Pendjari.

Access Porga is on main road from Ouagadougou and Fada N'Gourma (Upper Volta) to Natitingou, Tanguieta and Southern Dahomey. By plane from Cotonou to Natitingou.

Tracks in park 350 km (217 mi).

Special features Lakes with hippopotamus; hunting reserves outside the park; waterfalls of Tanougou near Batia on edge of park; country of the Somba people to the south of the park.

Open 20 December – 1 June.

Annual rainfall 1 050 mm (82 in); June–October.

Visitors 1969 About 1 000.

Principal mammal species:

Senegal galago c
Baboon a
Patas monkey c
Green monkey r
African hare r
Ground squirrel c
Cane rat r
Brush tailed porcupine r
Crested porcupine c
Common jackal r
Side-striped jackal r
Hunting dog r
Ratel c
Spotted hyaena c
Golden cat r (?)
Serval r
Caracal r
Lion c
Leopard r
Cheetah r
Aardvark c
Elephant r
Warthog a
Hippopotamus c
Bushbuck c
Buffalo a
Red-flanked duiker r
Crowned duiker r
Waterbuck a
Kob a
Reedbuck c
Roan antelope a
Korrigum c
Hartebeest a
Oribi c
Red-fronted gazelle r

Parc National du W du Niger

Locality In three countries: northern Dahomey, southeast Upper Volta and southwest Niger.

Area 11 020 sq km (4 250 sq mi). Dahomey 5 020 sq km (1 900 sq mi.) Upper Volta 3 000 sq km (1 140 sq mi.) Niger 3 000 sq km (1 140 sq mi.)

Formation 1954.

Vegetation Northern Guinea savanna with *Khaya, Pterocarpus, Parkia, Tamarindus, Daniellia, Buterospermum, Anogeissus, Acacia, Adansonia,* etc.; Gallery forest near rivers.

Special features Mekrou river, waterfall at 'Chutes de Koudou'.

Access From Kandi and Banikoara in Dahomey, Diapaga in Upper Volta, and

La Tapoa camp in Niger. Visas are not necessary when crossing state boundaries within the park.

Facilities No facilities in the park. Accommodation on boundaries of park at Diapaga (Upper Volta), La Tapoa Camp (Niger) and Banikoara (Dahomey). Forest post at Kéremou.

Open 20 December – 1 June.

Annual rainfall 700–900 mm (28–36 in); June–October.

Visitors 1969 About 200.

Principal mammal species:

Senegal galago c
Baboon a
Patas monkey c
Green monkey r
African hare r
Ground squirrel c
Cane rat r
Brush-tailed porcupine r
Crested porcupine c
Common jackal r
Side-striped jackal r
Hunting dog r
Ratel c
Spotted hyaena c
Golden cat r
Serval r
Caracal r
Lion c
Leopard r
Cheetah r
Aardvark c
Elephant c
Warthog r
Red river hog r
Hippopotamus c
Bushbuck c
Buffalo a
Red-flanked duiker r
Crowned duiker r
Waterbuck a
Kob a
Reedbuck c
Roan antelope a
Korrigum c
Hartebeest a
Oribi r
Red-fronted gazelle c

The Parc National de la Boucle de la Pendjari, and the Parc National de W du Niger are surrounded by extensive hunting zones, game reserves and partial game reserves. The parks and game reserves form a continuous area of about 25 600 sq km (10 300 sq mi).

Gambia

Abuko Nature Reserve

Locality 20 km (13 mi) from Bathurst, Gambia.

Area 62 ha (155 acres).

Formation 1968.

Vegetation Guinea savanna and riverain forest.

Special features Unique vegetation; animal orphanage; particularly good for birds.

Access By car from Bathurst, adjoins main road.

Facilities Two resthouses (for picnics only). No catering or overnight facilities.

Tracks None; footpaths only.

Open 15 November – 31 July.

Visitors 1969 3 000.

Annual rainfall 1 000 mm (40 in); July–October.

Principal mammal species:

Senegal galago c
Green monkey a
Red colobus a
Patas monkey a
Striped squirrel a
Sun squirrel a
Giant rat c
Cane rat c
Crested porcupine c
Brush-tailed porcupine c
White-tailed mongoose c
Marsh mongoose c
Gambian mongoose c
Slender mongoose r
Forest genet c
Blotched genet c
Civet c
Jackal c
Serval c
Leopard r
Bushbuck c
Red-flanked duiker r
Maxwell's duiker r
Crowned duiker r
Oribi r
Chimpanzee (introduced)

Ghana

Mole Game Reserve

Locality Northern Ghana, about 640 km (400 mi) north of Accra, near the town of Damongo.

Area 3 900 sq km (1 500 sq mi).

Formation 1958.

Vegetation Guinea savanna with *Burkea, Butyrospermum, Isoberlinia, Daniellia, Acacia*, etc. Gallery forest along rivers.

Special features Old settlements marked by *Anogeissus leiocarpus* and *Blighia sapida*.

Access By road from Tamale to Damongo; plane to Tamale.

Facilities Chalets, 15 beds, restaurant, swimming pool. Guides, vehicles.

Tracks 96 km (60 mi).

Open All year.

Visitors 1969 1 400.

Annual rainfall 1 075 mm (43 in); July–September.

Principal mammal species:

Green monkey a
Patas monkey a
Black colobus r
Baboon a
African hare c
Ground squirrel c
Cane rat a
Slender mongoose r
Kusimanse mongoose r
Civet r
Spotted hyaena r
Leopard r
Lion r
Aardvark r
Elephant r-c
Warthog c
Buffalo c
Bushbuck c
Red-flanked duiker c
Crowned duiker c
Waterbuck c
Kob c
Reedbuck r
Roan antelope c
Hartebeest a
Oribi c

Kujani Bush Game Reserve

Locality East of Ejura on the Kumasi-Atebubu road in central Ghana.

Area 280 sq km (80 sq mi).

Formation 1964.

Vegetation Woodland guinea savanna, riverine and relic forests.

Special features Intended as a research reserve.

Access By laterite road from Ejura on Kumasi-Atebubu road.

Facilities None.

Tracks None.

Open —

Principal mammal species:

Green monkey r
Lion r

Spot-nosed monkey r
Black colobus r
Patas monkey r
Baboon c
African hare c
Ground squirrel c
Cane rat r
Brush-tailed porcupine c
Forest genet r
Blotched genet r
Civet r
Slender mongoose r
Marsh mongoose c
Kusimanse mongoose c
Spotted hyaena r
Golden cat r
Leopard r
Aardvark r
Elephant r
Tree hyrax r
Red river hog c
Giant forest hog c
Warthog c
Buffalo c
Bushbuck c
Maxwell's duiker r
Yellow-backed duiker r
Crowned duiker r
Waterbuck r
Reedbuck c
Roan antelope r
Hartebeest r
Oribi r

Volta Game Reserve (proposed)

Locality On western shore of Volta Lake.

Area Approx. 2 600 sq km (1 000 sq mi).

Formation —

Vegetation Transitional between high forest and guinea savanna, fringing forest, swamps.

Special features Extensive shoreline of Volta lake; inlets of lake.

Access By boat from Akosombo; by road from Kumasi.

Facilities None. Base camp to be built on shore of lake.

Tracks None suitable for ordinary cars.

Open Not yet open to the public.

Annual rainfall 1 375 mm (55 in); March–October.

Principal mammal species:
Senegal galago
Potto
Green monkey
Spot-nosed monkey
Mona monkey
Red river hog
Warthog
Hippopotamus
Buffalo
Bushbuck

Black colobus
Patas monkey
Baboon
White-bellied pangolin
Crested porcupine
Brush-tailed porcupine
Golden cat
Hunting dog
Spotted hyaena
Leopard
Lion
Aardvark
Elephant
Tree hyrax
Manatee (in lake)
Bongo (?)
Red-flanked duiker
Bay duiker
Black duiker (?)
Yellow-backed duiker (?)
Maxwell's duiker
Crowned duiker
Waterbuck
Kob
Reedbuck
Roan antelope
Hartebeest
Royal antelope
Oribi

Ivory Coast

Parc National du Banco

Locality Close to Abidjan, Ivory Coast.

Area 30 sq km (11 sq mi).

Formation 1935.

Vegetation Primary rain forest.

Special features Comoé river, hills, Mont Boutourou.
forest trees.

Access By road from Abidjan (approx. 8 km or 5 mi).

Facilities No accommodation in the park because of closeness to Abidjan.
Guides available.

Tracks in reserve 60 km (37 mi); 30 km (19 mi) of the tracks are tarmac.

Open All year.

Annual rainfall No record; May–October.

Visitors 1969 About 1 000 per month.

Principal mammal species:
Chimpanzee r
Spot-nosed monkey r
Forest genet r
Giant forest hog r

Black colobus r
Civet r

Maxwell's duiker r
Bushbuck r

(Other species have probably disappeared due to heavy poaching.)

Parc National de la Comoé

Locality North-east Ivory Coast.

Area 11 500 sq km (4 270 sq mi).

Formation 1968 (previously a game reserve).

Vegetation Guinea savanna, gallery forests in the south.

Special features Comoé river, hills, Mont Boutourou.

Access By road Abidjan–Bouaké–Dabakala–Kong–Ouango-Fitini, 725 km (450 mi); or Abidjan–Abengourou–Bouna–Ouango-Fitini campement, 736 km (456 mi).

Facilities Campement-hotel with full accommodation at Ouango-Fitini in north of park. Airstrip. Forestry resthouse (11 beds) at Kakpin in south of park.

Tracks in reserve About 450 km (279 mi).

Open All year.

Annual rainfall 1 100 mm (44 in); April–October.

Visitors 1969 About 1 000.

Principal mammal species:
Baboon a
Black colobus c
Green monkey c
Lion c
Leopard c
Elephant a
Warthog c
Hippopotamus a
Buffalo a
Bushbuck c
Maxwell's duiker c
Red-flanked duiker c
Bay duiker r
Black duiker r
Crowned duiker c
Kob a
Waterbuck c
Reedbuck r
Roan antelope c
Hartebeest a
Oribi a

Parc National de la Marahoue

Locality Western Ivory Coast, northwest of Bouaflé.

Area 1 010 sq km (380 sq mi).

Formation 1968.

Vegetation Semi-deciduous forest, and wooded guinea savanna.

Special features Mont Seninlego 274 m (880 ft).

Access By road Abidjan–Bouaflé 378 km (234 mi); entrance into park at Gobazra 25 km (15 mi) from Bouaflé. Track to Mont Seninlego from Gobazra.

Facilities None in park.

Tracks in reserve 20 km (12mi) to Mont Seninlego.

Open January–May.

Annual rainfall No record; May–October.

Visitors No record.

Principal mammal species:

Chimpanzee a
Cane rat a
Spotted-necked otter r
Leopard r
Elephant c
Red river hog c
Warthog a
Hippopotamus c
Buffalo c
Bushbuck r
Bongo r
Kob r
Waterbuck r

Parc National du Mont Peko

Locality West of Daloa, western Ivory Coast.

Area 340 sq km (130 sq mi).

Formation 1968.

Vegetation Primary rain forest with the dominant species *Triplochiton sclexylon*, *Terminalia superba*, and *Funtumin latifolia*; large open areas.

Special features Mont Peko 1 004 m (3 212 ft) and Mont Kahoué 1 115 m (3 568 ft); forest animals and plants.

Access By road from Abidjan–Bouaflé–Daloa–Duékoué.

Facilities None. Accommodation available at Duékoué.

Tracks in reserve None, only footpaths.

Open All year.

Annual rainfall No record; June–October.

Visitors 1969 No record.

Principal mammal species:

Baboon a
Chimpanzee c
Giant pangolin r
Long-tailed pangolin c
Leopard c
Elephant a
Tree hyrax c
Warthog a
Hippopotamus a
Water chevrotain r
Buffalo a
Bushbuck
Bongo
Red-flanked duiker c
Maxwell's duiker c

Réserve de Faune D'Asagny

Locality West of Abidjan, Ivory Coast.

Area 300 sq km (104 sq mi).

Formation 1960.

Vegetation Savanna with borassus palms, islands of forest, extensive swamps.

Special features Rivers and lagoons.

Access By road from Abidjan–Dabou–Grand Lahou road.

Facilities None in the reserve.

Tracks in reserve None.

Open All year.

Annual rainfall No record; May to October.

Visitors 1969 No record.

Principal mammal species:

Black colobus r
Red colobus r
Chimpanzee c
Elephant c
Water chevrotain r
Buffalo a
Maxwell's duiker
Red-flanked duiker r

Réserve de Faune de Tai

Locality South-western Ivory Coast, close to the Ivory Coast–Liberia frontier, between the Cavally and Sassandra rivers.

Area 4 250 sq km (1 600 sq mi).

Formation 1956.

Vegetation Primary rain forest, patches of secondary forest on the edges close to abandoned villages.

Special features Rare forest mammals, large tall forest trees.

Access By tarmac road Abidjan–Yamoussokro–Bouaflé–Daloa–Duékoué–Guiglo 439 km (282 mi); graded road Guiglo–Tai 100 km (62 mi).

Facilities None in the reserve. Accommodation available at Duékoué and Guiglo.

Tracks in reserve None, except for some forest department footpaths. Part of reserve seen on Guiglo–Tai–Tabou road, and on Guiglo–Soubré road.

Open All year.

Annual rainfall No record; May–October.

Visitors 1969 No record.

Principal mammal species:

Red colobus a
Black colobus a
Diana monkey a
Spot-nosed monkey a
Chimpanzee a
Forest genet c
Civet c
Leopard c
Elephant a
Pigmy hippo r
Buffalo c
Bongo r
Bushbuck c
Bay duiker c
Striped duiker c
Maxwell's duiker a

Mali

Parc National de la Boucle du Baoulé

Locality About 105 km (65 mi) north-west of Bamako, Mali.

Area 5 430 sq km (2 174 sq mi).

Formation 1952.

Vegetation Sudan savanna, with *Butyrospermum, Daniellia, Lanea, Isoberlinia, Combretum* and *Pterocarpus*.

Special features ?

Access By road from Bamako.

Facilities Camps at Baoulé (12 beds) and Madina (4 beds). Guides.

Tracks in park ?

Open 1 November – 31 May.

Annual rainfall 1 100 mm (44 in); June–October.

Visitors 1969 400+.

Principal mammal species:

Chimpanzee
African hare
Side-striped jackal
Hunting dog
Spotted hyaena
Striped hyaena
Lion
Leopard
Elephant
Warthog
Hippopotamus
Giraffe
Buffalo
Bushbuck
Derby eland
Crowned duiker
Waterbuck
Reedbuck
Roan antelope
Hartebeest
Korrigum
Oribi
Dorcas gazelle

Réserve des Eléphants

Locality South of the Niger river 160 km (100 mi) west of Gao.

Area About 11 950 sq km (3 500 sq mi).

Formation 1959.

Vegetation Sudan and Sahel savanna, including *Acacia* spp. and *Adansonia*.

Special features ?

Access ?

Facilities None in the reserve.

Tracks in reserve ?

Open ?

Annual rainfall 400–600 mm (16–24 in).

Visitors 1969 ?

Principal mammal species:

Lion	Korrigum
Elephant	Dorcas gazelle

Réserve de Ansongo-Menaka

Locality North of the Niger river, between Ansongo and Menaka.

Area 17 500 sq km (6 650 sq mi).

Formation ?

Vegetation Sahel savanna, many *Acacia* spp.

Special features ?

Access ?

Facilities No facilities in the reserve; accommodation planned for 1972.

Tracks in reserve ?

Open 1 June – 31 October.

Annual rainfall 200–400 mm (8–16 in).

Visitors 1969 ?

Principal mammal species:

Striped hyaena	Scimitar-horned oryx
Lion	Addax
Cheetah	Dorcas gazelle
Manatee	Dama gazelle
Giraffe	Barbary sheep

Nigeria*

Borgu Game Reserve

Locality Between Niger river and Dahomey border, near to Kainji Lake.

Area 4 000 sq km (1 533 sq mi).

Formation 1963.

Vegetation Northern guinea savanna with patches of dense *Diosporus* forest. Fringing forest along Oli river; seasonal swamps.

Special features Oli river, eastern boundary on edge of Kainji Lake.

Access By road: Ibadan or Ilorin–Kishi–Kaiama–Wawa; or Ibadan–Jebba–Mokwa–New Bussa–Wawa. From northern Nigeria: Kotangora or Bida–Mokwa–New Bussa–Wawa. Entrance to park on Wawa to Yelwa road.

Facilities None in reserve. Accommodation at New Bussa, 40 km (25 mi) from reserve entrance.

Tracks in reserve About 80 km (50 mi).

Open All year.

Annual rainfall 1 075 mm (43 in); July–October.

Visitors 1969 About 2 000.

Principal mammal species:

Senegal galago a
Green monkey a
Patas monkey c
Baboon a
African hare r
Ground squirrel c
Sun squirrel r
Cane rat r
Crested porcupine r
Side-striped jackal r
Hunting dog r
Ratel r
Clawless otter r
Blotched genet c
Civet r
Gambian mongoose r
White-tailed mongoose r
Slender mongoose r
Egyptian mongoose r
Marsh mongoose r
Serval r
Caracal r
Lion r
Leopard r
Aardvark r
Elephant r
Rock hyrax r
Manatee (in lake) r
Warthog c
Hippopotamus c
Buffalo r
Bushbuck c
Red-flanked duiker c
Crowned duiker c
Waterbuck r
Kob a
Reedbuck r
Roan antelope c
Hartebeest c
Oribi c

Yankari Game Reserve

Locality 96 km (60 mi) south-east of Bauchi, North-east State, Nigeria.

Area 2 085 sq km (802 sq mi).

Formation 1955.

Vegetation Northern guinea savanna with *Pteleopsis, Combretum, Khaya, Tamarindus, Afzelia, Anogeissus, Burkea, Vitex,* etc.

Special features Gaji river and swamps, Wikki warm springs.

Access By road to Bauchi. Park road to Wikki camp from Bauchi–Yola road. Airport at Jos.

Facilities Full accommodation at Wikki camp; running water and electricity. Game viewing vehicles leave each morning and evening. Visitors can drive in their own cars.

Tracks in park About 640 km (400 mi).

Open 1 November – 30 June.

Annual rainfall About 1 000 mm (40 in); July–September.

Visitors 1969–70 About 2 000.

Principal mammal species:

Green monkey c
Patas monkey c
Baboon a
African hare r
Ground squirrel r
Crested porcupine c
Brush-tailed porcupine r
Civet r
Side-striped jackal r
Hunting dog r
Spotted hyaena ?
Lion r
Elephant c
Warthog a
Hippopotamus c
Buffalo a
Bushbuck c
Red-flanked duiker r
Crowned duiker c
Waterbuck a
Roan antelope c
Hartebeest a
Oribi c
Red-fronted gazelle r

Senegal

Parc National de Niokolo-Koba

Locality In eastern Senegal, north of the Guinea–Senegal border; about 640 km (400 mi) from Dakar.

* Another game reserve, the Upper Ogun Reserve, is being developed in the southern guinea savanna of Nigeria, about 80 km (50 mi) north of Ibadan.
The State governments of Nigeria are planning several new game reserves.

Area 8 130 sq km (3 100 sq mi).

Formation 1954.

Vegetation Guinea savanna of *Combretum, Parkia, Acacia,* etc.; grass plains; some areas of Sudan savanna.

Special features Tree hides, photographic hides.

Access By road or train to Tambacounda, and then south-east into park; by air direct into the park.

Facilities Air-conditioned hotel with full accommodation; two camps with running water and electricity, 200 beds, two restaurants; game-viewing vehicles, guides, etc.

Tracks in park About 800 km (500 mi).

Open 15 December – 15 June.

Annual rainfall 1 200 mm (48 in); August–October.

Visitors 1969 3 500.

Principal mammal species:

Senegal galago c
Red colobus r
Green monkey c
Patas monkey c
Baboon a
Chimpanzee r
Giant pangolin r
African hare a
Cane rat c
Crested porcupine c
Sun squirrel a
Ground squirrel a
Side-striped jackal c
Hunting dog a
Ratel r
Clawless otter r
Civet a
Forest genet c
Pseudogenet c
Palm civet r
Egyptian mongoose a
Slender mongoose c
White-tailed mongoose a
Gambian mongoose a
Spotted hyaena c
African wild cat c
Serval r
Caracal r
Leopard c
Lion c
Aardvark c
Elephant c
Rock hyrax r
Warthog a
Red river hog c
Hippopotamus ?
Bushbuck a
Derby eland r
Buffalo a
Red-flanked duiker a
Crowned duiker a
Waterbuck a
Kob a
Reedbuck c
Roan antelope c
Hartebeest c
Oribi c

Parc National de Basse-Casamanse

Locality Southern Senegal, near border with Portuguese Guinea.

Area 40 sq km (15 sq mi).

Formation 1970.

Vegetation Guinea savanna.

Special features Folk-lore camp, antelope enclosure, tree hides.

Access By road, 35 km from Ziguinchor.

Facilities Folk-lore camp.

Tracks 25 km (15 mi) and footpaths.

Open 1 December – 1 July.

Annual rainfall No record; August–October.

Principal mammal species:

Senegal galago a
Red colobus c
Green monkey c
Mona monkey c
Patas monkey c
African hare c
Sun squirrel a
Ground squirrel a
Forest genet c
Pseudogenet c
Palm civet c
Egyptian mongoose c
Marsh mongoose r
Leopard r
Aardvark r
Manatee r
Red river hog r
Hippopotamus r
Buffalo r
Bushbuck r
Yellow-backed duiker r
Maxwell's duiker c

Togo

Réserve de la Keran

Locality In north-eastern Togo, close to the Dahomey border.

Area 67 sq km (22 sq mi).

Formation 1950.

Vegetation Wooded savanna with gallery forest along streams.

Access About 450 km (280 mi) north of Lomé; by road from Lomé to Lama-Kara; difficult to reach.

Facilities None.

Tracks None.

Open All year.

Annual rainfall About 1 300 mm (52 in); July to September.

Principal mammal species:

Senegal galago
Black colobus
Baboon
Green monkey
Patas monkey
Tree squirrel
Ground squirrel
Giant rat
Cane rat
Crested porcupine
Brush-tailed porcupine
Civet
Genet
Palm civet
White-tailed mongoose
Slender mongoose
Marsh mongoose
Kusimanse mongoose
Side-striped jackal
Hunting dog
Spotted hyaena
Wild cat
Serval
Lion
Leopard
Aardvark
Elephant
Rock hyrax
Warthog
Red river hog
Hippopotamus
Buffalo
Bushbuck
Bongo
Red-flanked duiker
Crowned duiker
Waterbuck
Kob
Reedbuck
Roan antelope
Hartebeest
Korrigum
Oribi

Réserve de Malfacassa et Fazao

Locality About 320 km (200 mi) north of Lomé, close to Togo–Ghana border.

Area About 1 900 sq km (720 sq mi).

Formation 1951.

Vegetation Wooded savanna with dense trees in places, gallery forest by streams, thick woodlands on hills.

Special features Fazao mountains, many rivers especially the Mo and Anie.

Access Tarmac road for 320 km (200 mi) from Lomé to Blitta, followed by 25 km (15 mi) into reserve.

Facilities None in reserve.

Tracks About 40 km (25 mi).

Open All year.

Annual rainfall 1 200 mm (48 in); July–October.

Principal mammal species:

Senegal galago
Potto
Black colobus
Baboon
Patas monkey
Green monkey
Mona monkey
Striped and tree squirrels
Ground squirrel
Flying squirrel
Giant rat
Cane rat
Crested porcupine
Brush-tailed porcupine
Ratel
Civet
Genet
Palm civet
White-tailed mongoose
Slender mongoose
Marsh mongoose
Kusimanse mongoose
Jackal
Hunting dog
Spotted hyaena
Golden cat
Wild cat
Serval
Lion
Leopard
Aardvark
Elephant
Rock hyrax
Warthog
Red river hog
Giant forest hog ?
Hippopotamus
Buffalo
Bushbuck
Bongo
Red-flanked duiker
Maxwell's duiker
Bay duiker
Black duiker
Yellow-backed duiker
Crowned duiker
Waterbuck
Kob
Reedbuck
Roan antelope
Hartebeest
Oribi

Books for Further Reading

The following books and papers give further information on the large (and small) mammals of West Africa. Most of these quote other works which deal with West African mammals in greater detail.

For mammals which also occur in other parts of Africa, general books on African mammals may be consulted.

General

Allen, G. M. 'A Checklist of African Mammals,' *Bull. Mus. comp. Zool.* Harvard 83: 1–763, 1939.

Bigourdan, J. and Prunier, R. *Les Mammifères Sauvages de l'Ouest Africain et leur milieu*, Lechevalier, Paris, 1937.

Booth, A. H. *Small Mammals of West Africa*, Longmans, 1960.

Bourgain, P. *Animaux de Chasse d'Afrique*, La Toison d'Or, Paris, 1955.

Dekeyser, P. L. *Les mammifères de l'Afrique Noire Française*, I.F.A.N., Dakar, 1955.

Dorst, J. and Dandelot, P. *A Field Guide to the Larger Mammals of Africa*, Collins, London, 1970.

Happold, D. C. D. 'The distribution of large mammals in West Africa.' *Mammalia* 36(4), 1972

Roure, G. *Faune et Chasse en Afrique Orientale Française*, G. I. A, Dakar, 1956.

Webb, G. *A Guide to West African Mammals*, University of Ibadan Press, 1957.

Cameroon

Eisentraut, M. *Die Wirbeltiere des Kamerungebirges*, Hamburg, 1963.

Gromier, E. *La Vie des Animaux Sauvages du Cameroun*. Payot, Paris, 1937.

Jeannin, A. *Les Mammifères Sauvages du Cameroun.* Lechevalier, Paris, 1936.

Monard, A. 'Résultats de la mission zoologique: Suisse au Cameroun: Mammifères.' *Mem. I.F.A.N.* (*Centre Cameroon*), *ser. Sci. Nat.* 1: 13–57, 1951.

Perret, J. L. and Aellen, V. 'Mammifères du Cameroun de la collection J. L. Perret.' *Rev. Suisse Zool.* 63: 395–450, 1956.

Dahomey

Raynaud, J. and Georgy, G. *Nature et chasse au Dahomey*, Cotonou, Dahomey, 1969.

Ghana

Booth, A. H. 'Some Gold Coast mammals not included in Cansdale's *Provisional Checklist*,' *J. W. Afr. Sc. Assn.* 2: 137–138. 1956.

Cansdale, G. *Provisional Checklist of Gold Coast Mammals*, Govt. Printer, Accra, 1948.

Guinea

Cabrera, A. 'Catalogo descriptivo de los Mamiferos de la Guinea Espanola,' *Mem. Soc. esp. Hist. nat. Madrid* 16: 1–121, 1929.

Gromier, E. *La faune de Guinée*, Payot, Paris, 1936

Ivory Coast

Roure, G. *Animaux sauvages de Côte d'Ivoire et du versant atlantique de l'Afrique intertropicale*, Imprimerie Nationale, Côte d'Ivoire, 1962.

Liberia

Allen, G. M. and Coolidge, H. J. 'Mammals of Liberia,' in *The African Republic of Liberia and the Belgian Congo*, Vol. 2: 569–622. Harvard Univ. Press, 1930.
Kuhn, H-J. 'A Provisional Checklist of the Mammals of Liberia,' *Senck. biol.* 46: 321–340, 1965.

Nigeria

Rosevear, D. R. Many articles in the *Nigerian Field*, 1935–1950.
Rosevear, D. R. *Nigerian Mammals*, Govt. Printer, Lagos, 1951.
Rosevear, D. R. *Checklist and Atlas of Nigerian Mammals*, Govt. Printer, Lagos, 1953.

Senegal

Dekeyser, P. L. 'Le Parc National du Niokolo-Koba III: Mammifères,' *Mem. I.F.A.N.* 48: 35–77, 1956.
Dupuy, A. 'Le Parc National du Niokolo-Koba. XXXII: Mammifères (deuxième note),' *Mem. I.F.A.N.* 84: 443–460, 1969.

Togo

Baudenon, P. 'Notes sur les Bovides du Togo,' *Mammalia* 16: 49–61, 109–121. 1952.
Roure, G. *Animaux sauvages du Togo et de l'Afrique Occidentale*, Eaux et Fôrets, Lomé, 1966.

The following journals (besides others) contain papers on West African mammals:
Bulletin de l'Institut Fondamental d'Afrique Noire; Mémoires de l'Institut Fondamental d'Afrique Noire; Nigerian Field; Mammalia; African Wildlife.

Index

Species names in the section on National Parks and Game Reserves, pp. 78–100, are not included in the Index. Black and white illustration numbers are given in italics, colour illustration numbers are given in bold type.

ARCHANGEL

BY THE SAME AUTHOR

POETRY

In Doctor No's Garden

FICTION

Darien Dogs

Sandstorm

The Lost City

NON-FICTION

Sons of the Moon

Travels with my Trombone

Savage Pilgrims

ARCHANGEL

Henry Shukman

CAPE POETRY

Published by Jonathan Cape 2013
2 4 6 8 10 9 7 5 3

First published in Great Britain in 2013 by
Jonathan Cape
Random House, 20 Vauxhall Bridge Road,
London SW1V 2SA

www.randomhouse.co.uk

Addresses for companies within The Random House Group Limited can be found at:
www.randomhouse.co.uk/offices.htm

The Random House Group Limited Reg. No. 954009

A CIP catalogue record for this book
is available from the British Library

ISBN 9780224097420

The Random House Group Limited supports the Forest Stewardship Council® (FSC®), the leading international forest-certification organisation. Our books carrying the FSC label are printed on FSC®-certified paper. FSC is the only forest-certification scheme supported by the leading environmental organisations, including Greenpeace. Our paper precurement policy can be found at
www.randomhouse.co.uk/environment

Typeset in Bembo by Palimpsest Book Production Limited
Falkirk, Stirlingshire

Printed and bound in Great Britain by
MPG Printgroup, UK

for my father, Harold Shukman
1931–2012

CONTENTS

I

Nights

2004

FOUR A.M.

This is the hour the troubled man hears
the call of a train looping up a valley
and knows he must leave his home,
and also that he won't;

the hour the desperate wife
clutches her robe at the neck
and bathes herself in the light of a fridge,
having nowhere else to turn.

The poet looking out her window
at this hour sees she must
resolve her loves once and for all,
but only writes another poem.

Already a big dog lifts its *woof* into the air.
Something smaller answers: *yap yap*.
Soon the stars will withdraw one by one,
milk will lighten the coffee, and there won't

be anything left but ordinary day.
No one would guess not an hour ago
creation lay open like the back of a watch
and an early waker saw it all.

LIGHT AND DARK

All these years and I still don't understand
how it works, how the signal gets through
the bones of my hand, the bricks of this house,
the bank building opposite, and across miles

of suburb and field, pylons and roads,
hills and four rivers to precisely you,
in another city, another house, another room,
hunched by the bath with your phone in your hand,

sobbing. You can't bear to feel so split,
you gasp. Downstairs you hear a chair scrape,
and a man's voice. He laughs,
conversing with another ghost.

But I understand how light works.
Earlier your back gleamed like a guitar,
while the last leaves on the sycamore
flickered like a school of mackerel.

Later I will go out in a leopard-coat of light
without you: just me and the trees baring themselves
for winter, and the marbled paving stones,
and my empty hand shining.

READING CHEKHOV

Not until he was inches away
did he see how pale her lips were,
as if sea-bleached,
and how small and neat her teeth.
She was reading Chekhov to him,

the story about Anna and Gurov:
the two as lost in the world,
found in each other, as Adam and Eve
hand in hand at the gate.
And the cold Moscow street outside.

There's no time left for dishonesty,
she said. To be better,
we only need be who we are.
Why is that so hard? Why?
Her hair was everywhere

catching the light.
She couldn't get through a paragraph
without turning to him.
Their skins bloomed and she was fine
as a wishbone against him.

She closed the book.
It was silent in their world,
until a rustle, a lisp, a quiet unsticking
did away with any hope
of an answer.

THE BEGGAR FATHER

Your father
is country bound and comes to town no more.
He owns no bedding, rugs, or fleecy mantles,
but lies down, winter nights, among the slaves,
rolled in old cloaks for cover, near the embers.
Odyssey, XIV (trans. Fitzgerald)

Here he comes in his wet anorak,
ragged trouser-cuffs dark from gutters
after a night in the lanes, in the back
of a van, in a knocking shop,
reeking of tobacco and whisky.

Or up he clips with a spring in his step,
flightcase clutched to the chest,
the night of guilt already shaved away,
fragrant with Dove and Dior,
in and out for a family weekend.

The true prodigal is always the father,
they beat their sons to it every time.
With a stir of its stiff tail the old mutt
whimpers at the door. Then the cry
that is a laugh that is a kind of speech

rising from the heart of a boy
who runs across a sunlit yard
to the man who sways on his pins,
rubs his eyes, and sees and knows just enough
to open his arms and pull him in.

THE HOTEL

Cash, the sloe-eyed night-student wanted.
The walls were cardboard, the pillows pulp,
and the bathroom a plastic all-in-one,
the kind they fit on oil-rigs:

a knocking shop, and already home.
The ceiling bloomed with shadow,
the walls were bushes they pushed through
to their place of rest, fragrant with juniper.

God's ripe fruit: no matter how
they ate or drank it was not enough.
The wind whipped and stripped every leaf,
and they clung, two animals in a gale.

Something of them would always be there
now among the sinners and whores; even
when they'd torn that cheap place down,
and nothing was left but air.

LOVE SONGS

1. *Stop the Train*

It was the twins reunited,
spouses lost to one another,
Odysseus home to Penelope.

They burned from bar to bar.
She glowed in icon-light, happy.
Which made them sad: so short a time left.

A continent awaited them,
but they were being called away.
From the taxi she ran

as if to catch his train too.
When the doors screeched their warning
she held his face.

A whiff of nauseous diesel
floated through them.
The doors clunked, and there she was

beyond dust-speckled glass, still,
shocked, with her feet on another land.
And he pulling away.

2. *Blueberry Hill*

The first time, they resorted
to the park like teenagers.
They had just found the right
rhythm when there he was,
moaning in her neck.

She was too lovely, he said.
Traffic sucked by and he thought
of the train he had to catch.
Ten long minutes
they stayed, all they had.

His watch-face loomed
over a shoulder of hill,
stood still like the moon,
and shone
on what they had just done.

3. *Song of the Sea*

The world outside this room
has evaporated into cloud,
and from the cloud
again and again
comes the chime of the church clock.

This cabin is the chamber
of the seashell you sought as a girl,
with its endless inward spiral.
It whispered in your ear,
shell to shell.

Do you hear the sea stir
at the foot of this house?
That *dong-dong*, when it comes again,
is the bell of a buoy in high wind:
it's too late, love, we're far from shore.

4. *Hello Goodbye*

When you left at 4 a.m.
it was like the moon slipping its orbit
and drifting. No idea
where it was going or why.

It took fully half our time to say hello,
and half to say goodbye
and neither was nearly enough.

5. *She's Electric*

Sunday morning, a song chiming
from the kitchen window:
I sit on the wall in the sun
and hold your voice to my ear.
We need to take care of each other.
You lay out the rest of my years,
from this sunny wall
to the end of the garden.

6. *Jericho*

One more faultless autumn
evening spreads over the town:
a sky from Filippo Lippi,
a pair of low clouds
like two golden storks,
and the big silent jet
towing its luminous wake
through a final turn.

We haven't even talked of goodbye
but we're already broken.
Down the street the vicar
puts out his rubbish
in a big green bag,
a neighbour checks his oil.
I long for order, for the one
right act for just now.

Like an outcrop
over a lustrous Mediterranean
the clock tower ripens towards six,
when its hands will burn,
and once more, sounding
from the deep nave of the day,
the toll will tell us
our time has come.

7. *Here Comes the Night*

The world is full of omens –
glossy magpies, a train sliding
over a bridge as my taxi drives under,
even the slow stir of a birch

shredded by last night's wind
seems auspicious. And the twang
of ancient guitar on the radio
says we're still in love.

This morning the containers
in the gloom of a yard
are presents to unwrap, and the shadows
on meadows are arms to fall into.

Here it comes: we live on a rim
of light. Easy to forget,
on a dew-rinsed sun-washed morning,
how soon the night will come.

MARTYRS

When I sat on the side of the bed this morning
and thought of you, it was as if you were with me.

You were always a step ahead: yesterday
you'd said your body couldn't understand

our separating. You stood in the door,
cheek against the jamb, and frowned at me,

while I took in your face and eyes as they glazed red.
Didn't I realise? The house in flames around us.

HOUSEHOLD

The tons of brick and stone, the pounds of piping,
the sinks and china basins, the tiles and bathtub,
the hundredweights of board and joist,
the yards of flex and cable that wrap the house
like a net, the heavy glassed door, the rippled sheets
of window, bookcases, pictures in their frames,
tables, piano: what would it all weigh? One kiss,
one breathed declaration: the mass of love.

BEACH BY NIGHT

The breakers told them
who they were
a million years ago.
The longer they sat there
far from their homes,

the closer they came.
Call it love:
neither pleasant nor kind,
it cared little for them.
Any minute and they'd drop

into the same catastrophe
that beat the sea
against the shore,
and held the white pebble
in the sky, while

a burst of spray
travelled the pale line
of a breaker like the flame
of a fuse hurrying
towards the powder.

WHITSTABLE

None of it was the way we'd thought:
the shore path not even on the shore,
the oyster bar closed up, and the air
so thick the streetlamps bled through it.

I was running a fever. You worked
by the window, notebook in your lap,
then came to kiss me.

The king-size bed, the broad window
over the grey estuary: all ours.

We bundled up, ate chips on a bench
under a sky without a single star.

At the end, on the train, when I lay
in your lap, and you looked down
and I looked up, we saw the same future.

But we're in that future now,
and it's not what we imagined.

You're in your home, I'm in mine.
That day wasn't a promise,
it was just the day it was.

BOXING DAY

Here in America they don't have Boxing Day.
Just now you'll be sitting at the scarred table
in the low-beamed kitchen before a plate
of fruitcake, and a battlefield of crumbs.

There'll be tinsel on the dresser, while your son
picks raisins from wrapping paper as his grandma
asks a question that makes him pause and think;
then his answer that makes her pause and think,

and smile. Next door the fire's drawing breath.
And outside, the swaddling of cloudy sky,
the saturated lawn, the bank of dark trees,
as you head out for a walk. The hills will be iced

with snow, below their broad backs
they'll be black and brown, their streams splinters
of frosted glass. In your green cardigan
and yellow coat, your white hat pulled low,

and your heavy boots, you walk fast to keep warm,
your slender legs beating you uphill.
Your cheeks will be red, cold as ice-cream.
A man strides beside you, and perhaps high up

where the valley has settled like a model below
he holds you a moment, and you search inside,
find an alcove of warmth, go into it,
and as the snow on the high ground crunches

under your boots you think: yes, this is good.
This man you've loved a long time beside you,
and this country beneath you – the cottages
and hills, the lakes frozen into milk,

and the fruitcake, and shining coals in the grate –
you don't know what to call it but home.
There it is below: the valley
you've known better than any other.

Just now you're standing above your life.
You look west, south-west, across Easedale
to the Stickles and Ghills, where the sky
is like incense-smoke and candle-flame.

That's where I am. Over a sea, an island,
an ocean and mountains, looking back at you,
wondering if I can see you right from this distance,
a figure on the skyline of a hill,

walking again in your long strides.
I remember your cardigan's lanolin smell,
and the big pale mother's buttons,
and I think I can smell the smell of your flank

after a walk, like straw or hay, that filled
my head as you burrowed against me
with the cold tip of your nose. To miss someone:
what does it mean but think you see them still?

These years without you I'll keep my head down,
won't lift my eyes again to a brow far away,
where just now you've begun to sink
into the slope. I can still see your arms

swinging with your good pace,
but with every step you're coming down.
I can't see it but you must be on the path
that comes down Tongue by Grisedale.

Any minute and all that's left of you
will slip into the body of the hill.
Your waist, your chest, your shoulders,
the lovely ball of your head will be last to go.

The sun has gone out in your hair. Night
is coming on. It won't be long anyway
until the dark swallows you and you become
as see-through as the wind on the high ridge.

II

Archangel

THE SINGER

The day they got the Singer
my grandmother wouldn't touch it.
The clacking of its single tooth
terrified her, and the way it opened,
tipping over its hinge
so the bobbins, spikes and belts
of the gold and black carapace
almost fell in her lap.
She said it was the quality –
not like her own cross-stitch.
But Israel, my grandfather,
carried it up to the giant table
in the workshop where he'd sat
cross-legged like a buddha
for the past quarter-century,
and soon the others followed,
heaving Fristers, Rossmans
and more Singers up the narrow stairs
until the room clattered like a factory.
All day the sound drummed
down through the Soho tenement.
She'd eye the ceiling, while
her mad sister hid under the stairs.
But Dad, nine years old, curious,
slipped into the workshop
and sat in the twilight under the table,
among the lost needles and cigarette butts,
and the underside of the boards
gleaming over his head
like the belly of a stormcloud,
while the Singers thundered above.

THE TAILORS

Amkho scheer un ayren:
'the simple people of the scissors
and the ironing-board'

The last of them dead a decade
before I was born, they knew
the Englishman my father became,

in his tweeds and worsteds,
with his English wife on his arm,
voice polished up by college life,

but they died too soon to know his kids,
with our rowdy, ancestral noses
and sallow, unearthly dispositions.

What was it all for, their journey
of three hundred days, three thousand miles,
from Odessa, Warsaw, Lvov?

My grandfather was a man who
took off his jacket and cleared the kitchen
when he had to sign his name.

Here I am, all fingers and thumbs,
wishing I'd known them, wondering
who they were, stitching something

that never was – an absence
I come from, like the want
of a yarmulke over my scalp.

SHMATTAS

My father's favourite coat
was the one his brother made him
not long before he died.
Stiff corduroy, a quilted lining.

When he hugged us our cheeks
sank in, and the cloth released
a reek of stale cupboards,
as if the coat remembered

losses only our father knew.
We'd realise he too
had once been a child,
part of another family,

with his own parents, own siblings,
and we couldn't imagine
how they ever gave him up.
And he – how could he bear

to slip his arms into those sleeves,
silk-lined by his dead brother,
and plug the bone buttons in holes
his brother's fingers had sewn,

then stand at the door,
keys pocketed, pulling on
his gloves, and wait for us
to come and take his hands?

MOTHERLAND

He was a small man, solid
as pork, with a bayonet wound
from Mukden in his thigh.
He kept his head cropped
and wore a coat tailored
by his own hand; shaved
weekly at the shvitz;
in between, his stubble
glinted in the sun like coal.

She was a sylph in a scarf,
her thick dark hair tied back
from her white cheeks.
She cooked borshch and cabbage soup,
pickled pike in onions,
made kvass for the market.
Together they'd walk the Dnieper
at sunset, watching the ducks
leave their glassy ripples.

They married the day
the Commissar roused the troops
for another pogrom: Soho Jews
the rest of their lives, a little
English, a smattering of Polish,
and a fat river of Yiddish
that kept the broad Ukrainian fields,
the sudden woods and rivers
alive as an ache till they died.

What happens to the grass in spring
when you're not there?
It grows inside you, his brother said.
She looks out from the kitchen sink
through rain-pecked glass
at Soho roofs, at fire-escapes and brick,
at a formless sky of cloud or rain,
and feels the weather of Odessa
shift inside her, as a tube train
shudders up the line to Goodge Street.

SHOREDITCH

1917

Thursday night at the Bund:
London rain stings the black windows,
ten men in damp coats steam and smoke
beneath the hanging lamp.
Bialistok with the carmine cap
and Landsman whiskers
slaps down the *Standard.*
Now they're sending us to the trenches,
he screams in his high accent,
and reads out the headline:
If these brethren won't join up
we'll teach 'em the meaning
of anti-Semitism. Half Soho
has already gone to ground,
fled to Essex or the fens
or up to the northern munition works.
Wanted men, all of them,
even in this new land
that welcomed them
just a few years back.
Nathan sinks in his seat.
The meeting rattles on.
He uses one stub of nail
to clean another,
lights a Players and remembers
the zaddik of Lvov
who told him the secret archangels
Adriel and Barachiel
would watch over him
wherever he went.
Are they here now,
in the yellow fog under the ceiling?

All he can think of
is the thread he's run out of
for the Savile Row waistcoats.
If only we organizate,
Bialistok is shrieking.
Then they can't touch us.
We touch them. Oh yes, we touch –
Potok in blue corduroy
raps his pipe on the table.
May I remind you last time
we organizate? You want
this time in Bartholomews?
On and on they snipe.
Samael's talk, Amalek's work.
No peace or joy can come
from putting your trust in man.
His brother mutters in his ear:
The Winter Palace, the uprising.
We should be in Russia.
Nathan drops his chin to his chest
and searches for Adriel and Barachiel
in the floorboards between his shoes.
Don't forget: they will never desert you.

THE MAGGID OF CHERNOBYL

Over his razed shtetl, his abandoned lands,
the rabbi still keeps watch, they say:
Grand Reb Nachum of Chernobyl,
reaper of bitter wormwood,
sower of the Light of the Eyes,
who turned himself into a candle

for his followers. Once a year
the authorities let them in, the faithful
in their phylacteries and fedoras,
past the checkpoints and Geiger counters,
down the empty road
into the Exclusion Zone.

It's the old land now,
the *pushcha*, the deep forest
where the folktales were born,
land of ogres and mushrooms,
silent with hawk and duck,
and rivers of fleet-footed boar.

Past bushes alive with butterflies
they lift the lock on the shed
that is the old maggid's shrine,
a mile from the ruined power-station.
Quorum after quorum file through,
Kaddish and Tanya flowing

from their lips, and shockel
into the night, until all they see
is the light of the old rabbi's eyes.
Round his white sarcophagus
they burn so many candles
the eyes sting for the wax in the air.

SHIPPING OUT

1917

Back to Mother Russia, eh?
Our front not good enough?
Israel goes out
into the wordless, industrial
twilight on the Mersey wharf,
alien papers in hand,
coat pulled close,
and brother Nathan behind him,
silent and insubstantial
as a shadow.
Smoke from the funnels
leaks into the low cloud.
Up the gangway
of the Asiatic Line,
step by step they climb
towards the darkness
across the Baltic.

THE TRAIN

1917

The train's shadow swept
like a wing over the bright fields,
heading south. Thirty men
to a wagon, one latrine
and a barrel of sawdust.
It was October, still warm,
and Israel sat in the open door
smoking newspaper cigarettes.
The Revolution had come, you could feel it:
Kerensky, the Czar and his family
in a cellar near Omsk, the sky high
over the farmers clomping home
with their bullocks, over the labourers
drinking kvass by the hedges –
builders of the new society.
His brother Nathan cooked kasha
and potatoes at the leaky stove,
and the men solemnly thanked him.
He's a kind of holy man, your brother?
Israel couldn't say. He'd thought
the Hasidic nonsense forgotten,
they were socialists, that's why
they'd come back to Russia.
But while he couldn't help fretting
about the wives and children in London
Nathan just smiled, and his eyes shone
with a light from somewhere else.
He spent the days gazing down
the vanishing avenue of the rails,
while the autumn cold came on,
so the firs that lined the track
shone blue in the morning frosts.

TEPLUSHKI

2007

I wanted to look in the wagon
at the end of the army train
outside Archangel, but a soldier
on the hardcore waved me off.

Rough wood sides, a sliding door:
classic cross-country freight,
the kind my forebears rolled in
when the Army moved them south

just weeks before the Revolution
brought chaos to the railways,
to everything. Three years getting home:
every creaking mile in teplushki.

I glance in: dark wood, a partition,
an iron stove, but the soldier steers me off.
Nyet! he shouts. *Nein!* Then he grins.
Devushka: his CO's in there with a girl.

The drizzle comes back. We drive away,
through flat fields, shut-up villages,
towards a dream of colourless domes
you can just see across the Dvina.

CONSTANTINOPLE

1920

Three years out from London,
Nathan walks the dusk of Sultanahmet,
past the boarded-up synagogue,
up the hill to the old Roman church.
Swifts skim the domes.
Chestnut trees summon the coming dark.

Fat papierosi between his lips,
breastpocket full of army scrip,
he follows the advice of the old reb
from the shtetl of Lubartow
and climbs the leaning stairs

to a high balcony where they serve
small glasses of beer. He remembers
the reb stomping round his inn
in fur boots, filling up the drinkers
the way he filled his prayerhouse
with the holy fire of *hitlahavut*.

Nathan scans the Golden Horn:
a slip of blue Bosphorus,
and the island where the Mission
have him billeted – master tailor,
stitching Allied uniforms.

Russia is fleeing through the Dardanelles.
He and his brother have made it
all the way south, just ahead
of the tide of Civil War.
Vologda, Kursk, the Sea of Azov.

Thirty months of sleeping under tables,
earning their fares piece by piece,
while back in Shoreditch their wives
keep ahead of the rent, pulling
moonlight flits with the kids and barrows.

Stateless, paperless, he gazes
into the haze gathering
over the greatest city on earth,
sips his beer and admires the view
that just now, a thousand miles
from any home, seems all his own.

ARCHANGEL BIRCHES

2007

The black Russian birches
only turn silver forty feet up,
where their italic trunks
arc into the rainlike gauze
of their trembling world of leaf:
a line of them two hundred years old
at least, set to outlive us all,
shedding late-evening dusk
my ancestors would have known
that mid September in 1917
when they left the ship
in a wind cold off the Dvina.
That's all I find: between
the trees and the docks,
a deep tank of birch-light,
a green gloom lit with stars,
and the swaying proof
they once walked through it.

THE MAID

1920

Down the stairs onto the street,
he slips into the crowd and there she is,
stepping out of the dusk,
a creature made that instant
from the twilight, conjured like the swifts
from the air of Constantinople:

the one he is destined to find,
the one he's always been seeking.
Maid to a White countess:
an agate-eyed Jewess from Leipzig
with a husband beyond the Urals
and a stare that settles in his chest all week.

He recognises her, it stops him
in his tracks. But from where?
Sundays they meet, and they try
to find the answer, staying as long
as they can, eking out the daylight.
In the evening when he takes the steamer back,

a new man crossing a new sea,
he could reach out for the sun
and hold it in his hand like an apricot.
It's all just as the zaddik said:
this whole world is made of smoke,
and all that smoke is love.

ON PRINKIPO ISLAND

At night on the island
when they leave off
the clacking of dominoes
and sipping of apple tea,
and the casuarinas sigh
in the breeze off the Bosphorus,
he can hear the patter of paws
on the roof. Squirrels, that
scavenge the billet by night.

His brother in the next bunk
moans in his dreams.
Nathan can't sleep.
He slips into the grounds
and stares through the trees
at the gauze of moonlit sea.
She is with him, he knows it.
The whole world is here,
because God is here,
so she has to be.

He is too happy to think
of London. Or thinks of it
and sees no difficulty.
They will understand.
If God understands,
what man can fail to?
She is his bride, his *basherte*:
this *shidduch* was made
by Abraham himself.

He'll buy her a passage,
he'll bring her home.
He's in God's hands now.

A squirrel pauses
a few feet away,
and stares hard at him
with the black stone
of its eye.

HOTEL METROPOLE

Inside, the glow from the grate warms them
as they rest in the now-quiet bed.

This room must be a chamber of God's heart,
thinks Nathan – somewhere they'd live out

their days, if only they could.
He moves to the window.

Outside, the sky glows over a mosque,
and the rooftops spread as far as he can see,

panes here and there glinting like mica.
How small it is, this great city.

His happiness is a pressure,
as if the room and the silent streets

are not outside him; as if
this whole night and everything in it

is his own dream. Is this how God
repays his seeker: by giving him everything?

He stands at the window on chilled soles,
knowing now the real story was never his.

What relief. He and she are home,
the mission of Adriel and Barachiel accomplished:

they were children of the garden all along.
The night relaxes, sky and moon relax,

the angels rest too. This fire-lit room
is a garden that stretches to the end of time.

And all is as it should be – his love warm,
rustling in her sleep beside him.

THE HOUSE

On a hill above Beyoglu,
by an old church shaded with walnut trees –
the wooden mansion where she works.

His heart sinks as he sees it:
a quiet about it,
no laundry on the lines,

no maid sweeping, or sloshing out a pail,
no tradesman's horse amid cobbles of dung.
Not today.

He hammers on the door,
paces to the kitchen windows
at the back. Inside, it's all bare:

chandeliers draped in muslin,
furniture stacked against a wall.
He asks round the neighbourhood

in Yiddish, Polish, broken Turkish:
where have they gone?
No one knows. He waits all day

in the growing shadow
of the church. An old lady
gives him flatbread and a flask of water.

At dusk the house's windows turn clear,
and he sees into the blue depth
of deserted rooms. At night

a neighbour's light sends shadows
from the banisters
slanting up the walls.

At dawn a cat on the balcony
stares down with such large dark eyes
he jumps up, thinking it's her.

The neighbourhood wakes: infant cries,
the gush of a tap,
dogs calling from house to house.

He closes his eyes, searches the darkness within
for help. He offers all his fear
to God, but God won't take it.

He runs back to the hotel
where they would have met, had she come,
pays for a barrow, loads his boxes,

and sprints with his load like a madman
through the blind alleys of Terskaneh,
frantic for the docks.

A STREET IN KIEV

2007

God knows what they've been up to
but under the streetside awning
they bask as if in the light
of their own private sunset.

Her hair is a mess, shirt unbuttoned
to the clasp in the middle of her bra.
Her eyes shine like the small dark
aubergines they serve with drinks here.

The man slouches, legs crossed,
smooth as a billiard ball beside her,
an arm along her chair-back,
a gold chain winking

in the grey forest of his chest.
His fingers graze her back,
she arches, and he can feel
every one of her pores smile.

When a cloudburst opens
over the avenue, they look at one another
and don't even pull their chairs in
to keep out of the spray.

They sit there, stilled by the roar
of rain on the awning overhead,
that drums right through them,
rinsing every last cell of their bodies.

CROSSING THE BOSPHORUS

2008

He said we wouldn't find much
and we didn't. When the boat came
we stood side by side at the rail,
father and son, and it was good

to be out on the bright Bosphorus,
a breeze whipping up spray,
crossing a sea we knew
they had crossed ninety years before.

The thing was just to stay alive
if you could, he said, and gazed
at the distant green shore.
This is a good place.

Each spark on the waves winked
then went out: ten thousand pinpricks
piercing the cloth of errors
that was our history, there in the water.

THE DOCKS

1920

The cargo is loaded,
the passengers boarded,
bound for Marseilles.
Israel is on deck,
waiting for his brother.
Where are you, Nathan?
The ship lets out a bellow.
He feels the deck shudder
and the vessel start to move,
just as his brother races
into view on the quay.
But the gangplank is already up.
Israel waves, shouts across
the gun-blue water.
The huge cylinders are roaring,
and down on the docks
Nathan doesn't hear, doesn't see,
he wraps his scalp in his hands,
and the warehouses shift
behind him now, a shadowed
doorway swings into line
and swallows the dark figure.
Israel runs to the stern
but he can't see him,
and he never will,
and the ship is under steam,
making for the open sea
beyond the breakwaters.

REB IN FLIGHT

1970

Banking at 6,000 feet
the wing of the plane
winks above the aluminium sea,
over the glossy wake of boats
and hazy headlands.
The rabbi thinks of the Hasid's fire,
the flame one walks all day
without knowing it, the furnace
that is every moment,
the fire that is every act.
Forgive my blindness.
Or else take the scales
from my eyes, bring me home
on wings of light. They wheel
through a skin of cloud,
and land on burning ground,
on an evening strip
among unknown hills,
as he mutters his blessing
to the tyres and the air.

THE GOLDEN DRAGON

1975

It was raining in Soho.
We were in a Chinese round the corner
from where they used to live –
a street of tenements long since razed
for a wing of Middlesex Hospital –
when Dad told me how it ended.

Go on, he frowned behind his glasses
as I poked at a matchstick of ginger
resting in my strange Chinese spoon.
Eat it. You're twelve. The little sliver
exploded on my tongue with a freshness
I'd never known, like a sea breeze
carrying some far-off perfume.

The end of the story, you want? he asked,
and sucked up a dribbling chain
of noodles between his chopsticks.
*Who knows, boychick? They reached Russia
as the Revolution began, can you believe?*

His big eyes glared at me through his lenses.
*Chaos. A miracle either of them
ever got back – though only one did,
of course, and we both know which.
Otherwise I wouldn't be sitting here,
and nor would you.*

He chuckled and wiped his forehead.
Try this. I'd never had meat like it –
rich, hot, its sweet brown sauce steaming.
Nor had I ever known those people,

who had lived in another world,
whose London was not ours.
He came back alone, your grand-dad.
So what happened to his brother,
you want to know? I nodded,
my mouth full of sweet beef.

He shrugged. *Uncle Nathan.*
Not that I can tell you much.
For years they tried to get word.
Then they gave up. Hanya remarried,
to my uncle Toby the car-painter.

His eyes narrowed as he folded in
a hank of hot, wet greens.
There were more kids. You don't forget
but life picks you up, carries you.
The last person to see him alive
was my own dad, from the rail of a ship.
At least, that's what they thought.

He closed his hand round the little cup
of jasmine tea, and slurped from that next,
blowing steam from the surface.
I picked up my tea too, hot and fragrant,
and thought of a dark ship pounding north,
and two men in black coats out on deck.

But his eldest son, the rabbi, he flew
to Istanbul years later, just to see.
He was good with archives, city records,
and tracked down a white-haired Jew
running a nail stall in the Borsa.

Dad picked out a stick turning in his tea
and siphoned off a steaming mouthful.

Was it him? He couldn't be sure.
No papers, nothing. And the old guy
was half mad, sleeping at the back
of a rundown prayerhouse with a decrepit
old Hasid washed up from Yalta.

Two old shlemiels. They'd shockel,
dance and pray all night – all that.
Loudly too, the neighbourhood
was always complaining.

The waiter stacked one arm high
with our plates and bowls,
still fragrant from the meal,
and brought a saucer of orange quarters
with their bright rinds still on.
I bit into one, sweeter, juicier
than any orange at home.

The rabbi saw them one night.
There's a storm, the heavens open
right over the house, the Bosphorus flashing,
roofs lighting up, sky crackling
like a forest fire, and there they are,
two old guys stomping about
in the downpour, yelling their heads off,
strobe-lit by the storm.

So that's what you get. Crazies.
A couple of shlimazels out in the rain.
If he'd just caught that ship.
Still. Here we are, eh?
That's what matters, boychick.
Your grand-dad made it, so we made it.
You and me. We're here.

The door opened and a rain-chilled gust
blew in, along with the hiss of the street,
and chilled our legs under the table.
Then it closed, and once more
the warm hubbub settled around us,
and he signalled to the waiter for our bill.

SOTERIOLOGY

Adriel and Barachiel embrace him,
front and back, congratulate him
on his losses, and the old life shucks off
like a dead skin, all its loves and fears

gone in one blaze of grief.
His limbs won't stop trembling.
There's a desire in him
that can fulfill itself only in dovening.

Six days straight he neither eats nor sleeps,
but rocks in shivering prayer
through dawns and dusks,
until the angels have turned his heart

inside out. Now nothing
can touch him, not even death.
When he opens his eyes,
he blinks at a new world.

A zaddik, born twice,
he's a child of God no longer
in name but fact, who lives
neither in heaven nor on earth.

III

Days

MOSQUITO

What are you doing in these cold hills
far from warm nights and peeling hotel rooms?
You look as if you've fed on more than blood.
Do you miss the hot dark – a slick black river
sliding under breadfruit trees, a town alive
with skinny dogs and naked children,
and the song of millions of your kind?
You've learnt to lie low in these hostile hills,
where if a newspaper doesn't get you
the cold will. You hang there right where
you want to be, still as a crack in the wall.

SANTA FE TRAIL

I came over the rim and saw the distant dusty hills
risen from centuries of sun and wind,
lit like a lost Sahara by the first lick of dawn,
and thought of the wagon-trainers at this brow
hoping to see the golden city of holy faith,
and finding only more desert: empty.
There it is, see that stain, that's a garden
where they're growing chile, and that little
patch of dark, that's where the houses are.
The others aren't so sure. Coming down the hill
they meet the forerunners, offering stables, food.
A man with one tooth, stubbled, scrawny,
reeking of spirits, whose smile won't close,
whose black eyes shift and squint, says: *Sure*
you made it, this is Santa Fe, right here,
come on, a toast, you can stay with me,
I have room for all of you, and the mules.
He gestures to a forlorn compound
with a half-feathered hen and a hog
on a frayed rope. Can this be it? Santa Fe?
You know nothing but its name.
That was the old way: forerunners
got the innocents fresh from the trail.
Now we glide in in our own vessels,
and no one helps or misleads us.
Except who knows when they're getting close
to what they need, or what it feels like
to reach the journey's end? What traveller
can say for sure they ever arrive?

SANGRE DE CRISTO

The rag of lamb had twisted like a swami,
one leg right across the chest, another
behind the back, and the whole thing sunk,
losing whatever mass was once there
to the wind. But that wouldn't have been much:
it hadn't seen more than a day or two of life.
Already the fleece was gone on the limbs,
and honey-coloured bone showed through.

But the face, the long jaw laid on the grass,
the stretched neck, the firmly closed eyes –
there was such intention in it, like a human dreamer,
as if it had been caught trying to see something
or get somewhere, and now it had closed its eyes
to see something else, to see it better.

ST VINCENT'S AT NIGHT

The cannulas in his shrivelled arms
have bruised the ancient muscles black.
The cloudy afternoon has withered
to an overcast evening, but at the last minute
a weak glow lights the mountains.
He's talking about Venice long ago,
or is it Paris; there's New York too,
a bar where *all* the gays used to go,
and Ree proposed to her husband,
then the parties at their brownstone,
my *dear*, the things that went *on*.
His voice moves in and out of clarity
like a language I once spoke, while his eyes
roam the walls, and land on mine for punchlines.
He talks while he can. One knew *every*body,
now everybody's dead. The TV mutters
and the oxygen bubbles in its reservoir.
Like a cruise ship the edifice is calming
for the night, the visitors gone,
the day's quotas almost filled.
Of course he never *sat*isfied Ree
but that didn't matter, she had Lorenzo,
who all the gays would have *died* for a night with,
but nothing doing he *just* wasn't gay,
but *boy* could he mix a cocktail.
The gravel of cigarettes is in his throat.
Wherever home was – not another place
but time – it has closed like a blossom.
But its fragrance still fills every breath.
There's peace now that it's all too late:
the lovers, parties, drinks are gone
for good, there's nothing to be done
but listen to the murmur of the hospital
as it wheels through the night on its hill.

SPRING ON SAN ACACIO

Icicles weep, the gutters pour,
and before every window
a waterfall beats off our roof:
the whole house in tears.
When the sun breaks in,
the kitchen shivers, and we walk
from room to room blinking.
Clouds below us, the sun our peer:
this high new home
we have no name for.
In the long gleam of the mirror
we see only strangers.

POSTCARDS FROM NEW MEXICO

The mountains are icebergs
that have drifted out of childhood
to the rim of this night
to see if we remember them,
how big they are, how pale.

★

Where the hills fall
to the blue plain –
that's where the sea
should be: the coast,
the harbour of boats,
the traffic of nations.
But there's no sea
for a thousand miles.

★

A low coastal sun
finds the slit
between cloud and land
and makes every pine burn.

★

The base of the sky
is radiant with dusk,
and the stars
come out one by one,
then all at once.

★

Orion's lights
slip into
a neon dawn
and drown.

THE ADOBERO

The fall mornings are bitch cold,
there's frost on the studs set last week,
and the concrete footings are blue.

The Allsups coffee steams in your hand
as shadows drain from the hills.
Soon the sun will be lifting

from the canyon's lip.
The frame, ragged with electrics,
waits beside the barrow of skim.

All day you build up the old wall.
This land: its smoking galleons of cloud,
wind humming warm in the ear,

and the snow on the peaks
like loganberry icecream.
You wash off your trowel

in a bucket of sky, light up
a cigarette, and dream
of all the places you'll never go.

FIRESIDE

Midnight, the house at rest.
No one speaks, only the fire
muttering to itself
in its own happy sleep.

A woman and two men
stare into the grate.
There is no solution,
the love is cut all wrong.

Yet the night is full of grace:
light from the window
glowing on the frozen lawn,
the apple trees withdrawn
into their own stillness.

Even the complicated work
of love has dissolved
into shadow now,
thrown back, as perfect
and inevitable as the dawn.

SNOW ON CERRILLOS ROAD

Each car sheds its own aurora on the tarmac.
The trailer homes, shut up, no lights on,
bed cold under roofs the somnolent white of the sky.
Behind the big stores the desert is hoary.
Beneath the snow it will be the colour of night.

Our dark bus trembles down the highway.
At the intersection the red goes deep as midnight.
To the wipers' slow applause we turn
into another forecourt: Sleep Inn, Days Inn.
A half-lit lobby, a gleam of tile,
our indicator pulsing on the glass.

AutoZone and Dillards, Long John Silvers
and Home Depot, the Sonic and the Lamplighter –
the ballad of Cerrillos Road,
a serious business, this artery
of asphalt and rubber, the local life-support.

The bus splashes down the slow lane,
crunches through dykes of frozen slush.
Further in, the load of luggage lightened,
we bounce onto a street streaked white
as if someone had been cleaning a paint-brush.

The houses are hung with lights.
There's wood-smoke in the air,
sweaty, sweet, spiced with promises.
Cars hunched in driveways
bear dutiful loads on roof and hood.

Just the elderly couple and me now,
rumbling down a long wall to the last hotel.
Mightn't be this ain't the last run of the day,
says the driver, and whichever way
the negatives work, we know he's right.

A time like this, swamped by dark and snow,
you can't tell when your last chance might come.
The night is a swarm of flakes now,
flying thick and fast, until it's all
the wipers can do to keep things clear.

FARMER'S POEM

I used to write poems,
said the farmer at his stall
of small, sweet turnips.
But ten years back,
I lost my son. I waited
for him to go on.
His hazel eyes roamed
the resonant hall of the market –
the big windows steamy
from the shuffling crowd,
silver with drizzle outside –
then came to rest on mine:
clear, with a river depth
under their surface shine.
With stiff, swollen fingers
he brushed the dirt off a turnip,
and asked: *How many?*

OLD WAR

He left in the little car; I closed
my fingers over the back bumper
and held on. *I'm pushing you, Dad.*
And he let out the clutch. *Right-ho.*
Shadows flicked over the asphalt,
I smelled the blue smoke, my hands
tight on the chrome. The engine growled,
something wrong, too late,
I tripped, clung, dragged.

Dad stopped. The car panted, impatient.
His worsted knee on the tarmac.
I smelled his jacket's prickly weave.
A wail pierced the morning. He tried
to staunch it, keep it from the neighbourhood.
The trees behind their railings bloomed over us.
What happened? Shins a zigzag
blade of blood, knees a mess
of skin and grit. *I was holding on,* I cried.

THE NET

That time at dawn, no one else up,
I sent the ball down with the force
of someone I had not yet become
and never would, and it rang middle stump
like a bell, had it somersaulting into the net
where it hung by its point, while the ball
snuffled down the foot of the mesh.
And somewhere across the quiet early city,
a morning bell answered.

BLACK AND WHITE

for Sam

We woke to Bukowina lost in fog:
white hills, and black firs marching off
into the deep woods. While the parents
visited memorials, old villages, archives,

we sat in the Eagle Café, ordering
chorny-bialis – black-whites: a sludge
of hot harsh coffee under a pillow
of melting ice-cream, tall in a glass.

White-faced, black-haired, feeling serious,
we talked in the steamed-up window
of Poland, the ash and ovens, the birds
silenced, the smoke that hung

forever over the birches, the lost
relatives beyond counting. At dusk
we trudged like them – our ancestors –
through snow, back to the adults

in the boarding-house, past drifted
wagons and fences, startling at sounds
in the dark: the snort of a bullock,
the rasp and creak of a gate.

When the three figures loomed out
under a street-lamp, we flinched and ran.
There's safety in numbers, you breathed,
but we both knew that was never true.

THE PIPE

We grinned at each other in the dark
and took turns to suck. Orange Rough Rub,
the smallest tin they sold, the smoke
dense and sour, much too hot, but ours:
our pipe, our tin, our smoke.

This was what we were made for: to be out
in the city at night carrying contraband.
The little bowl scalded our palms,
the tin was like a grenade in my pocket:
proof the world would one day be ours.

LOVE LETTERS

What unerring instinct
carried us to the brown suitcase,
the wad of old envelopes,
the knot in the red string
so frayed it tore
with the slightest pull?

The full-stops had bled through,
blackening the paper with avowals,
making braille of the onionskin.
They had tousled the language,
forced things this way and that:
underwear, skin, hair degged
with sweat. *Garter. Arse.*

My sister's face went bright,
her nine-year-old eyes wide, lit
by a kind of joy or horror
in the dusk of the attic
with its one dusty skylight.
Our mum. Our dad.

Guilty, we tied them up
in the same red thread,
shut the clasps on the suitcase
still stale with the breath
of hotel nights from the fifties,
and let them be, left in search
of something we had no word for.

THE LONG DAYS

1. *Desert Nights*

When you arrived in your Chinese jacket –
your eyes their faded-jeans blue,
your hair neither blonde nor brown
but the colour of dust –

the mountains stepped back,
the desert invited us to tread
its glitter and crunch, and we rolled
into bed as if down a dune.
We had nothing to fear.

So we crossed the bridge into Mexico,
racked the margaritas on the bar,
the jukebox playing ranchera songs,
and chilled our throats and made our pledge:

bitter salt, sour lime, sweet liquor –
we would down it all.

2. *Black Mountain Drive*

That drive through the Black Range,
the grassland pale and sleek with dry spring,
the cottonwoods shining with new leaf –
and the Ford grumbling under our feet
found an old power that day, and roared
along the mountain road: remember?
You loved that lonely valley:
the bar with rooms at the back,
the podgy gallant cowboy
in from a ranch seventy mile off

who bought us the smoothest tequila –
This one's a sipper – then took you
for a chaste two-step. Your hair
lay long down your spine, and he touched
your back with his puffy hand.
Our quiet, our reticence said it all:
those dry mountains were just the start.

3. *Siena*

Remember the rain in Tuscany,
the English days that wrapped our honeymoon?
Autumn, and the stumpy vines were black,
the earth between them clotted like blood,
and the cypresses leaned into the weather.

That day on the Piazza del Campo –
its red-brick shell wide like a fan
that blew a breeze of grief at us.
What do you mean, you sobbed,
and I stared at the paper-thin skin
of your cheek and couldn't answer.

A gap the size of the piazza opened,
the ochre weave of its brick
as absolute as the sky.
How could you say that?
The rain came, and poured in ropes
into the vortex of the far gutter.

The umbrella drummed over our heads.
Tourists streamed across the square.
Waiters busied themselves with awnings,
couples abandoned their tables

to the wind, but we sat on,
while our coffees chilled
and we waited to see
who would first have the heart
to reach out and touch them.

4. *Teatime*

It was teatime when we arrived at the villa
under that grey October sky
across the camouflage of Tuscan hills.
The brick hall was hung with the owner's

Scottish ancestors, imperial figures
with guns and dogs, and wives in ballgowns.
In the cupboard, Typhoo and Marmite,
as if we'd brought with us the Britain we'd fled.

No wine, no cigarette, no tea or grappa
could warm us. All our cold history was there:
no escaping who we were or what
we'd done. *What is it?* you tried.

A man like me – steer clear of him.
If only I'd known. But I poured us wine,
unpeeled the prosciutto from the market,
and hoped we somehow could swallow it all.

HORSE AND RIDER, BY TER BORCH

The rider is so dog-tired the only thing
keeping him on his horse is exhaustion.
Whatever war he's dragged himself from
has changed everything; and this the one
respite, this stagger between troubles.
The mount, too tired to lift its head,
even lower it for fodder, could plant itself
like a table and sleep for days. Storm succeeds
to storm. The sky has wars of its own.

THE NIGHT SHE DIED

23 July 2011

Was her grand-dad a tailor?
Which shtetl did he come from?
Maybe his ship docked in Limehouse
like ours, and that's where they got
the name. Maybe there was
a god-loving reb in the line,
an ecstatic who'd found the key
to a celestial hall, and that's
where the music came from,
and the longing to be in God's palm,
instead of stumbling through cold daylight,
knocking from person to person,
from one frozen chance to another,
living on streets where no one
loves their neighbour, where even the lovers
know more of hate than love.

AFTER THE FUNERAL

Already, gaps among the cars
in the carpark.
What can we do
but show up, then roll away
to the rest of the afternoon?

We come to pay something,
a little time, and to see
what nothing looks like,
then return to our day
among the quick –

to sunshine, to dry grass
bright in the short
autumn afternoon,
and to mountains dusty-green
on the skyline,

covered in pines that live
much longer than us,
and to wind just now
making the stalks and fronds
tremble along the verge.

MOTHER

This morning in her third sleep
or fourth, a final chance to catch up

on the broken night of children,
there's a murmur in each breath,

a friction in her throat.
Perhaps she feels it, and rolls over.

Then the silence she slips back into:
it's not that you daren't move

for disturbing her, you couldn't if you wanted.
Sleep fills the room like the morning light.

ODYSSEY

for Saul

The almost-four-year-old's
gleaming body seems to sink
into the ground beyond the path's brow
as he walks off: walks off
again, another rehearsal.
One day it will be years,
not minutes, before he reappears
with something precious.
What leave-taking does he practise?
What journey is already
in the making? We are all
figures on the skyline
of a hill, we are all
disappearing out of the light.

LOVERS

for Stevie

He drapes an arm across my chest.
Daddy: a sigh, and he's still.

I turn to see if he's asleep
and I can creep away.

His eyes gleam at me
in the dark like two wells.

The little fingers tighten
round my wrist.

One night years from now
he'll lie like this with a lover.

What could this night
have to do with the other?

EVE

What a night it was:
the Big Dipper tilted in the sky,
last week's snow thickened
by tonight's frost, the bells
clanging across the meadow,
and the whole house redolent

of the wine they were mulling –
uncle, mother, cousins, aunts –
the cloves, orange, sugar by the mug
steaming and steeping the old stone walls.
And in the living room the dark tree
rising from a loam of gifts.

No one was out but me. Stars prickled
and the old chestnut held so still.
I heard them call my name – a flurry
of calls that died away – and didn't move
from my station by the hoary rosemary
brittle as a bunch of sparklers.

I wasn't waiting for anything,
only for the feel of waiting
to rise from my feet, now welded
to the hard lawn, up to my head.
It was a night when anything
that might happen would happen.

Again they called me.
Like this, I could love them all.
Smoke rose into an effervescent sky.
In the village a dog's bark
lost its edge in the cold.
And still I kept my station.

KING JOHN'S LANE

for Clare

A tree has come down in the beech woods,
fractured white among the ground ivy

in the almost-dusk where we're walking.
This marriage – it was always a step ahead

of us, knowing something we didn't:
the losses we'd one day forget,

the promises we hadn't yet seen.
It still is. That's why we couldn't leave it.

We had what the Indians look for,
a perfect polarity in our stars: a strong horse

of a union, they say, built to last.
So on we go down the old lane,

with its sheen of worn leather,
past the dead boughs and the living,

towards a vanishing-point we can't see
in the murk, that won't stop unfurling

in verges of nettle and dock,
and the path freely giving itself to us.

NOTES & ACKNOWLEDGEMENTS

Archangel

During the First World War, there were 65,000 Jewish tailors living between the West and East End of London. Few were British citizens; when conscription was introduced in 1916, the authorities and a vitriolic press debated whether or not, as friendly aliens, they should be sent to the Western Front. In the end they were offered a choice: either the British trenches or deportation back to Russia and the Tsar's Army. In the late summer of 1917, several thousand Jewish tailors, among them my grandfather and great-uncle, found themselves disembarking in Archangel, in northern Russia, just as the Revolution was starting.

In the ensuing chaos the Tsarist Army fell apart, and these would-be recruits were issued with All-Russia rail passes. Some took their chances east, hoping to cross Siberia and sail for California; a few went north, wintering for unknown reasons in the Arctic; most went south, funneling down to the Black Sea in the hope of crossing southern Europe back to their families in Britain. Many were never heard of again.

The section 'Archangel' is deeply indebted to *War or Revolution*, the only historical account of this little-known episode of the First World War, written by my late father, the historian Harold Shukman.

Shmattas

shmattas: a term for clothing, often associated with the Jewish ragtrade.

Motherland

shvitz: literally, sweat: a traditional Jewish steam-bath similar to Russian or Turkish baths.

kvass: a fermented drink made from rye, popular in Russia.

Shoreditch
zaddik: a Hasidic or more generally Jewish holy man.
Adriel and *Barachiel*: two 'unknown' archangels, beyond the familiar seven.
Samael: an angel of death, a destroyer.
Amalek: the first and arch-enemy of the Israelites when they reached Canaan; also, the psychic and cosmic nemesis of all that is pure and godly in man and creation.

The Maggid of Chernobyl
phylacteries: the *teffilin* traditionally worn by Orthodox Jews.
Tanya: a revered text written by Shneur Zalman of Liadi (1745–1812), an early rabbi of the Hasidic movement.
shockel: to rock in prayer.

The Train
kasha: a buckwheat porridge.

Constantinople
papierosi: Russian cigarette with cardboard filter.
Lubartow: a village in south-eastern Poland.
hitlahavut: a condition of grace associated in Hasidism with clear vision of the eternal fire that burns within everything.
Mission: The British military mission that helped to administer Constantinople immediately after the First World War.

Archangel Birches
Dvina: the largest of Archangel's several rivers.

On Prinkipo Island
basherte: a beloved, or bride, with whom a man has a special affinity in a formal marriage arrangement.
shidduch: a formal betrothal organised by a matchmaker.

The Golden Dragon
shlemiel, *shlimazel*: a fool.

Soteriology
dovening: rocking in prayer (as in shockeling).

Grateful acknowledgement to the editors of the following:
Cimarron Review, *Guardian*, *London Review of Books*, *New Republic*, *New Welsh Review*, *New Writing 15*, *Times Literary Supplement*.

'Light and Dark' appeared in the anthology *Answering Back* (ed. Carol Ann Duffy).

Thanks to Neil Rollinson, Robin Robertson, Sam Willetts, Bill Broyles, Tony Hoagland; and, once again, to Clare Dunne.

LOW PURINE DIET FOR BEGINNERS

Easy Delicious Recipes and Tips for Managing Gout and Joint Pain to improve Health & Reduce Uric Acid Levels.

Anderson Pinedo

Anderson Pinedo

ISBN:

Printed in the United States of America

Disclaimer
This publication is designed to provide competent and reliable information regarding the subject covered. However, the views expressed in this publication are those of the author alone, and should not be taken as expert instruction or professional advice. The reader is responsible for his or her actions. The author hereby disclaims any responsibility or liability whatsoever that is incurred from the use or application of the contents of this publication by the purchaser of the reader. The purchaser or reader is hereby responsible for his or her actions.

ABOUT THE AUTHOR

Mr. Anderson Pinedo is a highly skilled and informed nutritionist who is passionate about assisting individuals in achieving maximum health and well-being via tailored dietary plans and fitness routines. With over ten years of expertise,

he has assisted over 500 people of different ages and backgrounds in achieving their health objectives and living better, more satisfying lives. Mr. Pinedo has a bachelor's degree in nutrition from a renowned institution, as well as many continuing education courses and certifications in nutrition, exercise science, and wellness coaching.

He has a thorough awareness of the special health demands and issues that elders confront.

Mr. Pinedo has worked with people from various walks of life, including athletes, retirees, and persons with chronic health concerns, throughout his career.

He is well-known for his compassionate and individualized approach to nutrition counseling,

as well as his ability to create practical and successful nutrition regimens that are tailored to each individual's specific requirements and lifestyle.

TABLE OF CONTENTS

INTRODUCTION

A diet plan that limits the consumption of foods high in purines is known as a low-purine diet. The body converts purines, which are organic compounds present in many meals, into uric acid. Uric acid is typically eliminated through the urine, but if the body creates too much of it or cannot adequately eliminate it, it can accumulate and crystallize in the joints, leading to excruciating gout episodes. For those with gout, kidney stones, or even other illnesses that result in excessive amounts of uric acid in the body, a reduced purine diet is often advised. A low-purine diet can help lower the risk of gout episodes, stop the development of kidney stones, and enhance general health.

Limiting the consumption of purine-rich meals and increasing the consumption of low-purine foods is the major objective of a low-purine diet. Organ meats, shellfish, alcohol, and certain vegetables like spinach as well as asparagus are among the foods high in purines. Fruits, vegetables, cereals, and especially low-fat dairy products are examples of low-purine foods.

A low-protein diet is not always the same as a low-purine diet. Protein is a necessary food that is required for sustaining healthy muscular mass. Those who consume low-purine foods should instead concentrate on obtaining their protein from lean meats, and poultry, including plant-based sources such as beans and tofu.

Those who consume a reduced purine diet might additionally have to alter their lifestyles to maintain their health. They can include drinking plenty of water, working out frequently, and controlling your stress.

Generally, gout and other illnesses associated with excessive amounts of uric acid in one's system can be effectively managed with a reduced purine diet. It's crucial to collaborate with a medical professional or a qualified dietitian to create a custom low-purine diet plan that suits your needs and tastes.

CHAPTER ONE

UNDERSTANDING PURINES

What are Purines?

Purines are occurring natural chemicals present in a variety of foods and bodily cells. These are fundamental components of DNA, RNA, and ATP (adenosine triphosphate), all of which are required for the normal operation of body cells.

When the body breaks down purines, uric acid is formed. Uric acid is generally eliminated in the urine by the kidneys. Nevertheless, if the body creates too much uric acid or the kidneys have no way to adequately eliminate it, it can accumulate in the blood and lead to hyperuricemia.

Foods High In Purine

Some foods that have a lot of purines are:

Organ meats: liver, kidneys, tripe, and sweetbreads

There are three game lumps of meat: deer, rabbit, and pheasant.

Anchovies, sardines, mackerel, herring, scallops, mussels, and shrimp are examples of seafood.

Meat drippings and sauces

Beer and other alcoholic beverages, including beer and spirits in particular.

How Purines Impact the Human Body

When we ingest purine-containing meals, our bodies convert them into uric acid. Uric acid is generally eliminated by the kidneys in urine; however, if the body creates too much uric acid or if the kidneys are unable to remove it correctly, it can accumulate in the blood and cause hyperuricemia.

Hyperuricemia can cause a buildup of urate crystals inside the joints, which could also activate the immune system and induce inflammation, resulting in gout. Gout is a kind of arthritis that can produce sudden and severe joint pain, inflammation, and redness. Over time, gout episodes can cause joint damage and raise the chance of acquiring other diseases, including kidney stones as well as cardiovascular disease.

Moreover, purines might harm the kidneys. Urate crystals can develop in the kidneys when uric acid levels are excessive, leading to the production of kidney stones. High amounts of uric acid can cause kidney damage and raise the chance of developing a progressive kidney condition over time.

In addition to nutrition, additional variables such as genetics, age, gender, and medicines can impact purine levels in the body. Consequently, it is essential to consult with a healthcare expert or a qualified dietitian to identify the optimal nutritional approach for controlling hyperuricemia, gout, and other associated health issues.

CHAPTER TWO

LOW PURINE DIET FOR BEGINNERS

Significance for Beginners

A decrease in inflammation and protection against the development of some illnesses has been related to a reduction in the consumption of foods high in purines.

Benefits of a Low Purine Diet for Beginners

With a reduced purine diet, you cut back on purine-rich foods. Foods produced from animals, such as meat, fish, and poultry, and even some plants, naturally contain purines. Uric acid is created when the body breaks down purines, and too much of it can lead to gout and kidney stones.

There are some benefits to adopting a low-purine diet.

Prevents Gout Flare-ups and Attacks:

Those who are prone to gout might benefit from following a low-purine diet to prevent painful attacks. This is because gout is caused by an excess of uric acid in the body and reducing uric acid levels in the blood may be achieved by restricting the consumption of foods high in purines.

Benefits of Kidney Stone Treatment:

Those who eat a low-purine diet are less likely to develop kidney stones. High urine uric acid levels are linked to the development of kidney stones, making it imperative to control one's intake of purines.

Enhances Cardiovascular Health:

There may be a link between eating fewer purines and improved heart health. In addition to having a high purine content, red meat, and organ meats are also heavy in saturated fats, which have been associated with both high cholesterol and an increased risk of cardiovascular disease. The heart can be protected by limiting the consumption of certain foods.

Aids in Achieving and Keeping a Healthy Weight:

It has been hypothesized that following a low-purine diet might help people shed extra pounds. Cutting less on high-calorie and high-purine meals will help you reach and keep your ideal weight.

Supports Healthy Cartilage and Bone and Prevents Joint Damage:

Joint health and mobility can be significantly impacted by inflammation in the body, however, eating fewer purines may help. Gout is a kind of arthritis defined by the deposition of uric acid crystals inside the afflicted regions, causing pain and swelling in the joints. Joint inflammation and pain experienced by gout sufferers may subside if they refrain from eating foods rich in purines.

Facilitates Healthy Digestion:

In addition, a purine-restricted diet can improve gastrointestinal health by promoting the consumption of fiber-rich foods such as fruits, vegetables, and whole grains. Constipation and other digestive issues might be avoided with frequent consumption of these foods.

Inhibits Inflammation:

Following the advice given before, eating fewer purines can help lower systemic inflammation. Chronic inflammation has been related to a wide variety of diseases, including cardiovascular disease, cancer, and even autoimmune disorders. A decrease in inflammation and protection against the development of some illnesses has been related to a reduction in the consumption of foods high in purines.

Boosts Health in Every Way:

Because of its emphasis on whole, natural foods high in nutrients, the low-purine diet is often considered to be beneficial to health. Changing to a low-purine diet may improve one's health in many ways, including increased energy, better mood, and reduced chance of acquiring chronic diseases. With a reduced purine diet, you cut back on purine-rich foods. Foods produced from animals, such as meat, fish, and poultry, and even some plants, naturally contain purines. Uric acid is created when the body breaks down purines, and too much of it can lead to gout and kidney stones.

CHAPTER THREE

GETTING STARTED ON A LOW-PURINE DIET FOR BEGINNERS

Shopping List for Beginners on Low Purine Diet

It's crucial to organize your meals and food shopping if you're on a low-purine diet. A thorough shopping list for a low-purine diet is provided below:

Proteins:

Chicken (breast and legs) (breast and legs)

Turkey (breast) (breast)

Eggs

dairy products with less fat (milk, yogurt, cheese)

Beans in Tofu (kidney beans, navy beans, lentils)

seeds and nuts (almonds, walnuts, chia seeds, flaxseeds)

Almond butter (in moderation)

Fruits:

Berries and Apples (strawberries, blueberries, raspberries)

Grapes, Cherries, and Melons (cantaloupe, honeydew)

Oranges

Pineapple

Watermelon

Vegetables:

Asparagus

Broccoli

Carrots

Cauliflower

Celery

Cucumber

Eggplant

the beans

Kale

Lettuce (romaine, iceberg) (romaine, iceberg)

Peppers (red, green, yellow) (red, green, yellow)

Potatoes (white, sweet) (white, sweet)

Spinach

Tomatoes

Zucchini

Grains:

Dark rice

pasta, whole wheat bread, and quinoa

Oils and fats:

Almond oil

rapeseed oil

Avocado lard

seeds and nuts (almonds, walnuts, chia seeds, flaxseeds)

Almond butter (in moderation)

salad dressing with less fat

Beverages:

Liquids Herbal tea

Fruit juice with little sugar

reduced-fat milk

Foods to Eat on a Low Purine Diet:

- Milk, yogurt, and cheese are low in purines and high in calcium, which can help reduce gout attacks. Low-fat dairy products.
- Fruits: Fruits rich in antioxidants and vitamins yet low in purines include cherries, strawberries, blueberries, bananas, and grapes.
- Vegetables: Vegetables strong in nutrients and low in purines include kale, spinach, broccoli, cauliflower, carrots, and potatoes.
- Grains: Whole grains are high in fiber and low in purines, such as brown rice, quinoa, and oats.
- Eggs: Eggs have a high protein content and a low purine content.

- Nuts and seeds: Nuts and seeds, such as almonds, walnuts, and flaxseeds, are rich in healthful fats yet low in purines.
- Oils: Oils high in healthy fats and low in purines include olive oil, canola oil, and avocado oil.
- Drinks: Water, herbal tea, and low-sugar fruit juice are all purine-free and can aid in the body's removal of uric acid.

Foods to Avoid on a Low Purine Diet:

- Organ meats: Since they are high in purines, organ meats including liver, kidney, and sweetbreads should be avoided.
- Seafood: Because they are high in purines, seafood such as anchovies, sardines, mussels, scallops, and herring should be avoided.
- Meat: Because it contains a lot of purines, meats like beef, hog, and lamb should only be eaten seldom.
- Beer in particular has a high purine content and can cause gout episodes.
- Sugary beverages: Sugary drinks contain a lot of fructose, which can raise uric acid levels in

the blood. Examples of these drinks are soda and fruit juice.

- High-fructose corn syrup: This sweetener, which is frequently present in processed foods, raises blood levels of uric acid.
- Fried foods: Fried foods should be avoided since they are high in purines.

CHAPTER FOUR

MEAL PLANNING ON A LOW-PURINE DIET FOR BEGINNERS

21-Day Meal Plan

Day 1

Greek yogurt with fresh berries and almonds for breakfast

Quinoa and black bean salad with mixed greens for lunch

Baked lemon herb salmon with roasted asparagus for dinner

Day 2

Breakfast: omelet with spinach and mushrooms served with whole wheat bread

Grilled chicken breast with mixed veggies for lunch

Dinner: lentil soup with a salad on the side

Day 3

Oatmeal with fresh fruit and almonds for breakfast

Lunch: Stir-fried turkey and vegetables with brown rice

Baked chicken breast with roasted sweet potatoes and green beans for dinner

Day 4

Smoothie with Greek yogurt, fresh berries, and spinach for breakfast

Lunch: stuffed quinoa bell peppers with a side salad

Baked lemon herb tilapia with roasted Brussels sprouts for dinner

Day 5

Cottage cheese with fresh fruit and almonds for breakfast

Lentil and vegetable soup with a side salad for lunch

Grilled chicken breast with roasted asparagus and sweet potatoes for dinner

Day 6

Breakfast: omelet with spinach and mushrooms served with whole wheat bread

Baked lemon herb salmon with mixed veggies for lunch

Dinner: Stir-fry lentils and vegetables with brown rice

Day 7

Smoothie with Greek yogurt, fresh berries, and spinach for breakfast

Quinoa and black bean salad with mixed greens for lunch

Roasted chicken breast with roasted Brussels sprouts for dinner

Day 8

Greek yogurt with fresh berries and almonds for breakfast

Lentil and vegetable stir-fry with brown rice for lunch

Baked lemon herb tilapia with roasted sweet potatoes for dinner

Day 9

Breakfast: omelet with spinach and mushrooms served with whole wheat bread

Grilled chicken breast with mixed veggies for lunch

Dinner: stuffed quinoa bell peppers with a side salad

Day 10

Oatmeal with fresh fruit and almonds for breakfast

Lentil and vegetable soup with a side salad for lunch

Roasted chicken breast with roasted asparagus and sweet potatoes for dinner

Day 11

Smoothie with Greek yogurt, fresh berries, and spinach for breakfast

Quinoa and black bean salad with mixed greens for lunch

Baked lemon herb salmon with roasted Brussels sprouts for dinner

Day 12

Cottage cheese with fresh fruit and almonds for breakfast

Grilled chicken breast with roasted asparagus and sweet potatoes for lunch

Dinner: lentil soup with a salad on the side

Day 13

Breakfast: omelet with spinach and mushrooms served with whole wheat bread

Baked lemon herb tilapia with mixed veggies for lunch

Dinner: stuffed quinoa bell peppers with a side salad

Day 14

Smoothie with Greek yogurt, fresh berries, and spinach for breakfast

Lentil and vegetable soup with a side salad for lunch

Roasted chicken breast with roasted Brussels sprouts for dinner

Day 15

Greek yogurt with fresh berries and almonds for breakfast

Quinoa and black bean salad with mixed greens for lunch

Baked lemon herb salmon with roasted asparagus for dinner

Day 16

Breakfast: omelet with spinach and mushrooms served with whole wheat bread

Grilled chicken breast with mixed veggies for lunch

Dinner: Stir-fry lentils and vegetables with brown rice

Day 17

Oatmeal with fresh fruit and almonds for breakfast

Lentil soup with a side salad for lunch

Roasted chicken breast with roasted sweet potatoes and green beans for dinner

Day 18

Smoothie with Greek yogurt, fresh berries, and spinach for breakfast

Lunch: stuffed quinoa bell peppers with a side salad

Baked lemon herb tilapia with roasted Brussels sprouts for dinner

Day 19

Cottage cheese with fresh fruit and almonds for breakfast

Lentil and vegetable stir-fry with brown rice for lunch

Grilled chicken breast with roasted asparagus and sweet potatoes for dinner

Day 20

Breakfast: omelet with spinach and mushrooms served with whole wheat bread

Quinoa and black bean salad with mixed greens for lunch

Dinner: lentil soup with a salad on the side

Day 21

Smoothie with Greek yogurt, fresh berries, and spinach for breakfast

Baked lemon herb salmon with mixed veggies for lunch

Dinner: stuffed quinoa bell peppers with a side salad

Delicious Recipes for Beginners on Low Purine Diet

Spinach and Mushroom Omelet

Ingredients:

2 huge eggs

1 cup Sliced mushrooms

½ cup Fresh spinach leaves

¼ cup onions, diced

1 tbsp Olive oil

pepper and salt as desired

Directions:

In a nonstick skillet over medium heat, warm the olive oil.

Add the onions and cook for two minutes, or until soft.

Cook the mushrooms and spinach for two to three minutes, or until wilted.

Add salt and pepper to beaten eggs in a bowl.

The egg mixture should cook in the skillet for two to three minutes or until set.

Banana Oatmeal

Ingredients:

1 cup Oats, rolled

1 cup of low-fat milk or water

1 mashed ripe banana

½ teaspoon of cinnamon

1 teaspoon of honey (optional)

Directions:

In a small pot, combine the oats, milk or water, banana puree, and cinnamon. Bring to a boil.

Reduce heat, cover, and simmer for 5-7 minutes, stirring occasionally, until the oatmeal is cooked and thickened.

If desired, drizzle with honey when serving hot.

Avocado Toast with Smoked Salmon

Ingredients:

1 slice of bread with low purines

¼ mashed avocado

1-ounce Smoked salmon

1/8 cup lemon juice

pepper and salt as desired

Directions:

The bread should be lightly toasted.

Avocado should be mashed with salt, pepper, and lemon juice.

On the toast, spread the avocado mixture, then add the smoked salmon.

Serve right away.

Yogurt and Berry Parfait

Ingredients:

½ cup of Plain low-fat yogurt

¼ cup of fresh berries (such as blueberries, raspberries, or strawberries)

1 teaspoon of honey

2 tablespoons Granola, (optional)

Directions:

In a glass or container, arrange yogurt, berries, and granola (if using).

Add honey to the dish.

Offer cold.

Tofu Scramble

Ingredients:

½ brick crumbled firm tofu

¼ cup of onions, diced

¼ cup of bell peppers, diced

¼ cup of mushrooms, chopped

1 tbsp Olive oil

pepper and salt as desired

Directions:

In a nonstick skillet over medium heat, warm the olive oil.

Add the onions and simmer for two minutes, or until tender.

Add the bell peppers and mushrooms, and simmer for 2 to 3 minutes, or until tender.

Stir the veggies and crushed tofu together in the skillet.

Add salt and pepper to taste and simmer for a further 2 to 3 minutes, or until well cooked.

Serve warm.

Quinoa Breakfast Bowl

Ingredients:

2 cups of mixed vegetables and four ounces of fried chicken breast

½ cup of cherry tomatoes

¼ cup of cucumbers, diced

¼ cup Red onions, cut

Balsamic vinaigrette, two teaspoons

Directions:

Put a mixture of greens on a platter.

Add cherry tomatoes, cucumbers, red onions, and grilled chicken as garnishes.

Add a balsamic vinaigrette drizzle.

Offer cold.

Green Smoothie Bowl

Ingredients:

1 cup of washed and drained dry lentils

1 cup carrots, diced

½ cup celery, chopped

2 cloves of minced garlic

½ cup of diced onions

4 cups of vegetable broth low in salt

1 tbsp Olive oil

pepper and salt as desired

Directions:

Over medium heat, warm some olive oil in a saucepan.

Add the onions and garlic, and simmer for two to three minutes, or until tender.

Cook the carrots and celery for a further two to three minutes.

Bring to a boil the lentils and vegetable stock in a saucepan.

Lower heat, cover, and simmer for 20–25 minutes or until lentils are done.

To taste, add salt and pepper to the food.

Serve warm.

Fish with Asparagus Grilled

Ingredients:

4 ounces of Grilled salmon

½ pound trimmed of asparagus

Olive oil, 1 tbsp

pepper and salt as desired

Directions:

Grill at a medium-high temperature.

Olive oil should be used to coat the asparagus before salt and pepper.

Grill asparagus for 5 to 7 minutes, or until fork-tender and faintly browned.

Grilled salmon can be seasoned to taste with salt and pepper.

Serve warm.

Eggplant and Zucchini Rollatini

Ingredients:

1 medium eggplant, cut crosswise

1 medium zucchini, cut crosswise.

½ cup Ricotta cheese

¼ cup grated parmesan cheese

¼ cup finely minced fresh basil

½ cup Low-sodium tomato sauce

pepper and salt as desired

Directions:

Turn on the 375°F oven.

Slices of eggplant and zucchini should be grilled or roasted until soft, about 5-7 minutes.

Combine ricotta cheese, parmesan cheese, and finely chopped basil in a bowl. To taste, add salt and pepper to the food.

Each slice of eggplant and zucchini should have 1 tablespoon of tomato sauce on it.

Each slice should have a teaspoon of the ricotta mixture added to one end before rolling.

The leftover tomato sauce should be poured over the rolls after placing them in a baking dish.

Bake for 15 to 20 minutes, or until well heated.

Serve warm.

Turkey Lettuce Wraps

Ingredients:

4 ounces Lean ground turkey

¼ cup red bell peppers chopped

¼ cup water chestnuts, diced

1 tbsp Low-sodium soy sauce

1 teaspoon of ginger, grated

4-6 substantial lettuce leaves

Directions:

A nonstick skillet should be heated to medium-high.

Add the ground turkey and simmer for 5 to 7 minutes, or until browned and well cooked.

For an additional 2-3 minutes, add red bell pepper, water chestnuts, soy sauce, and ginger.

Pour the mixture onto lettuce leaves and wrap them up. Offer cold.

Roasted Vegetable and Quinoa Salad

Ingredients:

1 cup cooked quinoa

1 cup of roasted veggies (such as zucchini, bell peppers, and onions)

¼ cup feta cheese crumbles

2 teaspoons freshly chopped parsley

1 teaspoon of lemon juice

1 tbsp Olive oil

pepper and salt as desired

Directions:

Combine cooked quinoa, roasted veggies, feta cheese, and parsley in a dish.

To create the dressing, combine the lemon juice, olive oil, salt, and pepper in another bowl.

Next, add the dressing and toss to coat the quinoa and veggie combination.

at room temperature or chilly.

Tuna and Avocado Salad

Ingredients:

4 ounces of drained canned tuna

sliced avocados, half-ripe

¼ cup red onion, chopped

¼ cup of celery, chopped

1 teaspoon Lemon juice

1 tablespoon olive oil

pepper and salt as desired

Directions:

Combine tuna, avocado, red onion, celery, and parsley in a bowl.

In a separate dish, mix lemon juice, olive oil, salt, and pepper to produce a dressing.

Toss the tuna and avocado mixed with the dressing to coat.

Offer cold.

Sweet Potato and Black Bean Burrito Bowl

Ingredients:

1 diced medium sweet potato

½ cup of rinsed and drained black beans

¼ cup red onion, diced

¼ cup bell pepper, diced

2 teaspoons freshly chopped cilantro

1 tbsp Olive oil

½ teaspoon cumin powder

pepper and salt as desired

Directions:

Set the oven to 400 °F.

Spread the sweet potatoes in a single layer on a baking sheet after tossing them with olive oil, cumin, salt, and pepper.

Sweet potatoes should be roasted for 20 to 25 minutes, or until they are soft and gently browned.

Roasted sweet potatoes, black beans, red onion, bell pepper, and cilantro should all be combined in a dish.

at room temperature or chilly.

Grilled Vegetable Sandwich

Ingredients:

½ medium zucchini cut lengthwise

½ medium eggplant cut lengthwise

¼ cup of red onion slices

2 tablespoons of tomato sauce low in sodium

2 pieces of whole-wheat bread

1 tbsp Olive oil

pepper and salt as desired

Directions:

Grill or grill pan should be heated to medium-high.

Olive oil should be brushed on zucchini, eggplant, and red onion. Salt and pepper should also be used.

Grill veggies for about 5-7 minutes, or until they are soft and mildly browned.

Cauliflower Fried Rice

Ingredients:

1 head of grated cauliflower

2 minced garlic cloves

½ cup onion, chopped

carrots, diced, in a cup

½ cup bell peppers, diced

¼ cup Low-sodium soy sauce

2 teaspoons Olive oil

pepper and salt as desired

Directions:

Over medium-high heat, warm up the olive oil in a big skillet.

Cook the onion and garlic together for about 3–4 minutes, or until the onion is transparent.

Cook the bell peppers and carrots for a further 3–4 minutes, or until they start to soften.

In a pan, combine soy sauce and grated cauliflower; simmer for 5 to 7 minutes, or until the cauliflower is soft.

To taste, add salt and pepper to the food.

Serve warm.

Lentil and Vegetable Stew

Ingredients:

1 cup Dried green lentils

1 cup of kale, chopped

1 cup of carrots, chopped

1 cup of celery, diced

1 cup of onion, chopped

4 cups low-sodium vegetable broth

2 teaspoons of olive oil

1 tablespoon of dried thyme

pepper and salt as desired

Directions:

In a big saucepan, warm up the olive oil over medium heat.

Add the onion and simmer for 3 to 4 minutes, or until transparent.

Cook the carrots and celery for a further 3 to 4 minutes, or until they begin to soften.

Bring to a boil the lentils, vegetable stock, and thyme in the pot.

The lentils should be soft after 20 to 25 minutes of simmering over low heat.

Kale is added and cooked for a further 5-7 minutes, or until wilted and tender.

To taste, add salt and pepper to the food.

Serve warm.

Grilled Chicken and Vegetable Skewers

Ingredients:

4 ounces of cubed, skinless, and boneless chicken breast

½ Sliced medium zucchini

½ medium red onion

½ cup of cherry tomatoes

¼ cup Low-sodium Italian dressing

pepper and salt as desired

Directions:

Grill or grill pan should be heated to medium-high.

On skewers, arrange chicken, zucchini, onion, and cherry tomatoes.

Add salt and pepper after rubbing with Italian dressing.

Skewers should be grilled on the grill for 8 to 10 minutes, rotating a few times, or until the chicken is fully cooked and the veggies have a light sear.

Serve warm.

Roasted Vegetable and Brown Rice Bowl

Ingredients:

1 cup of brown rice, cooked

one cup of roasted veggies (such as Brussels sprouts, sweet potatoes, and beets)

¼ cup goat cheese crumbles

2 teaspoons freshly chopped parsley

1 teaspoon of lemon juice

1 tbsp Olive oil

pepper and salt as desired

Directions:

Cooked brown rice, roasted veggies, goat cheese, and parsley should all be combined in a dish.

To create the dressing, combine the lemon juice, olive oil, salt, and pepper in another bowl.

The rice and vegetable combination will be coated once the dressing has been added.

Whether hot or cool, serve.

Baked Lemon Herb Salmon

Ingredients:

4-ounce salmon fillets

2 teaspoons Olive oil

2 tablespoons freshly chopped parsley,

1 tablespoon freshly chopped thyme

1 tablespoon of lemon juice

1 tablespoon lemon zest

pepper and salt as desired

Directions:

Turn on the 375°F oven.

Olive oil, parsley, thyme, lemon zest, lemon juice, salt, and pepper should all be combined in a small bowl.

Salmon fillets should be placed on a baking pan covered with parchment paper.

Salmon fillets should be covered in the herb-lemon mixture.

Bake the salmon for 12 to 15 minutes in a preheated oven, or until it is well-cooked.

Serve warm.

Quinoa Stuffed Bell Peppers

Ingredients:

4 medium-sized peppers

1 cup cooked quinoa

½ cup rinsed and drained low-sodium canned black beans

½ cup of chopped zucchini

½ cup of minced onion

½ cup tomato dice

¼ cup finely minced fresh parsley

2 teaspoons Olive oil

pepper and salt as desired

Directions:

Turn on the 375°F oven.

The bell peppers' tops should be cut off, and the seeds and membranes should be removed.

Cooked quinoa, black beans, zucchini, onions, tomatoes, parsley, olive oil, salt, and pepper should all be combined in a bowl.

Place the bell peppers on a baking sheet that has been lined with parchment paper after stuffing them with the mixture.

Bake for 30-35 minutes in a preheated oven, or until the filling is well-cooked and the peppers are soft.

Serve warm.

Chicken and Vegetable Stir Fry

Ingredients:

4 oz. of skinless, boneless, and thinly sliced chicken breast

½ cup sliced carrots

½ cup broccoli, chopped

½ cup Chopped bell peppers

¼ cup Low-sodium soy sauce

2 teaspoons Olive oil

2 minced garlic cloves

pepper and salt as desired

Directions:

In a wok or large skillet, heat the olive oil over high heat.

Add the chicken and simmer for an additional 3 to 4 minutes, or until cooked through.

For an additional one to two minutes, add the garlic.

Cook the bell peppers, broccoli, and carrots in the pan for 3 to 4 minutes, or until the veggies are just starting to soften.

Stir the chicken and veggies as you pour the soy sauce over them.

Continue to simmer for a further 5-7 minutes, or until the veggies are soft and the chicken is well cooked.

To taste, add salt and pepper to the food.

Serve warm.

Spaghetti Squash with Turkey Meatballs

Ingredients:

1 medium-sized spaghetti squash

8 ounces of lean turkey meat

2 cups of chopped mushrooms

1 cup of chopped onion

½ cup grated Parmesan cheese

½ cup chopped fresh parsley

2 teaspoons Olive oil

1 teaspoon Oregano, dry

1 teaspoon of powdered garlic

pepper and salt as desired

Directions:

Turn on the 375°F oven.

Slice the spaghetti squash lengthwise, then remove the seeds and membranes.

Put the cut-side-down squash halves on a baking sheet covered with parchment paper.

Roast the squash in the preheated oven for 30 to 40 minutes, or until it's

CHAPTER FIVE

LIFESTYLE CHANGES TO SUPPORT BEGINNERS ON LOW PURINE DIET

Importance for Beginners

Implementing lifestyle adjustments can help to promote a reduced purine diet and reduce the likelihood of gout attacks. Here are a few ideas:

Keep Hydrated:

Consuming enough water as well as other fluids can aid in the removal of uric acid from the body and the prevention of kidney stones.

Restrict your Alcohol Consumption:

Because alcohol, particularly beer, and spirits, can raise uric acid levels in the body, it is advisable to restrict or avoid it.

Regular Exercise:

Daily exercise can help you keep a healthy weight, enhance your circulation, and lower your chance of gout episodes.

Stress Management:

Because stress can induce gout episodes, it's critical to discover strategies to control stress, such as via yoga or meditation.

Keep a Healthy Weight:

Being overweight increases the risk of gout episodes, thus maintaining a healthy weight with a balanced diet including frequent exercise is essential.

Get Adequate Rest:

Obtaining adequate sleep is vital for general health, as well as reducing stress and lowering the likelihood of gout episodes.

CHAPTER SIX

SUPPLEMENTS FOR SENIORS ON A GOUT DIET

While a well-balanced diet is the best method to get all of the nutrients you need, several supplements can help those on a reduced-purine diet. The following substances may be beneficial:

Vitamin C:

Vitamin C is an antioxidant that could also help lower uric acid levels in the blood. Lowering inflammation can also assist to avoid gout episodes. Citrus fruits, strawberries, kiwi, and especially bell peppers are good sources of vitamin C, but a supplement may be essential for people who do not obtain enough from their diet.

Omega-3 Essential Fatty Acids:

Omega-3 fatty acids have been shown to help decrease inflammation in the body but may assist to avoid gout episodes. Omega-3 fatty acid-rich foods include fatty fish such as salmon and sardines, in addition to flaxseeds and chia seeds.

Those who do not obtain enough fish oil from their diet may benefit from a fish oil supplement.

Magnesium:

Magnesium can aid in the prevention of kidney stones, which are a consequence of gout. Spinach, almonds, and black beans are good sources of magnesium, but a supplement may be essential for people who do not obtain enough from their diet.

Vitamin B-Complex:

B-complex vitamins can aid in general health and energy generation. Whole grains, almonds, and leafy green vegetables are good sources of B vitamins, but a supplement may be essential for people who do not obtain enough from their diet.

Probiotics:

Probiotics can assist to enhance gut health, which is crucial for people with gout since gut health affects uric acid absorption and excretion. Fermented foods like yogurt, kefir, and kimchi are good sources of probiotics, but a supplement may be essential for people who don't receive enough from their diet.

Curcumin:

Curcumin is a turmeric component that has anti-inflammatory effects. Those suffering from gout may find it beneficial, as inflammation can contribute to gout episodes. Those who do not obtain enough turmeric or curcumin from their diet may benefit from taking a turmeric or curcumin supplement.

Vitamin D:

Vitamin D is essential for bone health as well as immunological function. Fortified foods like milk and cereal, as well as fatty seafood like salmon and tuna, are good sources of vitamin D. Those who do not obtain enough from their food or through sun exposure may need to take a supplement.

CONCLUSION

Those with gout, kidney stones, or even other illnesses associated with excessive levels of uric acid in the blood may benefit from a reduced purine diet. The diet emphasizes fruits, vegetables, whole grains, and low-fat dairy products while restricting or eliminating items rich in purines, such as red meat, organic meat, and seafood.

Meal preparation and a slow but steady approach can help make the switch to a reduced purine diet less of a shock to the system. Consulting a licensed dietician might be beneficial as they can offer individualized guidance and assistance.

A low purine diet and general health can be supported by making adjustments to one's lifestyle, such as drinking enough water, keeping a healthy weight, including engaging in frequent physical activity.

Low-purine diets can benefit from taking supplements, but it's best to check with a doctor first because supplements can interfere with medicines and have additional negative effects.

When it comes to controlling gout and other illnesses caused by elevated uric acid levels in the blood, a reduced purine diet is often recommended. For newcomers, with the correct encouragement and direction, it may be a long-term lifestyle shift that improves their health and well-being.

Printed in Great Britain
by Amazon

28976510R00036